METHODS IN MOLECULAR BIOLOGY

Series Editor
John M. Walker
School of Life and Medical Sciences
University of Hertfordshire
Hatfield, Hertfordshire, UK

For further volumes:
http://www.springer.com/series/7651

For over 35 years, biological scientists have come to rely on the research protocols and methodologies in the critically acclaimed *Methods in Molecular Biology* series. The series was the first to introduce the step-by-step protocols approach that has become the standard in all biomedical protocol publishing. Each protocol is provided in readily-reproducible step-by-step fashion, opening with an introductory overview, a list of the materials and reagents needed to complete the experiment, and followed by a detailed procedure that is supported with a helpful notes section offering tips and tricks of the trade as well as troubleshooting advice. These hallmark features were introduced by series editor Dr. John Walker and constitute the key ingredient in each and every volume of the *Methods in Molecular Biology* series. Tested and trusted, comprehensive and reliable, all protocols from the series are indexed in PubMed.

Photorespiration

Methods and Protocols

Edited by

Berkley J. Walker

Plant Research Laboratory and Department of Plant Biology, Michigan State University, East Lansing, MI, USA

Humana Press

Editor
Berkley J. Walker
Plant Research Laboratory
and Department of Plant Biology
Michigan State University
East Lansing, MI, USA

ISSN 1064-3745 ISSN 1940-6029 (electronic)
Methods in Molecular Biology
ISBN 978-1-0716-3801-9 ISBN 978-1-0716-3802-6 (eBook)
https://doi.org/10.1007/978-1-0716-3802-6

This Humana imprint is published by the registered company Springer Science+Business Media, LLC, part of Springer Nature.
The registered company address is: 1 New York Plaza, New York, NY 10004, U.S.A.

Paper in this product is recyclable.

Preface

Photorespiration has long fascinated researchers interested in photosynthesis and central metabolism since its discovery and characterization over the last 50 years. Some of this interest likely comes from its apparent contradictions. On the one hand, it consumes a large portion of energy from the light reactions and releases previously fixed carbon at high rates; on the other hand, it appears to be absolutely required in higher plants to deal with the inhibitory products of rubisco oxygenation reaction. Additionally, there is growing evidence that it is linked to many important aspects of plant metabolism including nitrate assimilation, one-carbon metabolism, amino acid synthesis, and energy metabolism. Photorespiration also attracts interest due to the nexus of techniques that were required for its discovery, initial characterization, and now subsequent research. The substrate of photorespiration (2-phosphoglycolate) was first associated with photosynthesis in the radiolabeling experiments resolving the C3 cycle. Classical "bucket biochemistry" (requiring buckets of plant material to purify the requisite rubisco) resolved the substrates for 2-phosphoglycolate production. Early gas exchange experiments described a light-dependent uptake of CO_2, and the pioneering use of forward genetics in *Arabidopsis* resolved the biochemical pathway of photorespiration. While all these techniques (or modern variations) of them are still used to study photorespiration, new tools are required to ask questions not previously asked. These techniques include bioinformatical approaches, mass spectrometry, NMR, and computer modeling. These diverse methods mean that those interested in researching photorespiration must constantly hone their skillsets to continue to understand this vital pathway in plants. Hopefully, this book gives others the tools to do so!

East Lansing, MI, USA *Berkley J. Walker*

Contents

Contributors

CYRIL ABADIE • *Institut de Recherche en Horticulture et Semences, Université d'Angers, Beaucouzé, France*
DEBARATI BASU • *Foray Biosciences, Cambridge, MA, USA*
FEDERICA BRANDIZZI • *MSU-DOE Plant Research Laboratory, Michigan State University, East Lansing, MI, USA; Department of Plant Biology, Michigan State University, East Lansing, MI, USA; DOE-Great Lakes Bioenergy Research Center, Michigan State University, East Lansing, MI, USA*
ANDREA BRÄUTIGAM • *Computational Biology, Faculty of Biology, CeBiTec, Bielefeld University, Bielefeld, Germany*
CASEY BUTLER • *Department of Plant Pathology and Crop Physiology, Louisiana State University, Baton Rouge, LA, USA*
MAIKE COSSE • *Department of Plant Cell Biology, Botanical Institute, Christian-Albrechts University Kiel, Kiel, Germany*
QIANWEN DENG • *College of Agriculture and Biotechnology & ZJU-Hangzhou Global Scientific and Technological Innovation Center, Zhejiang University, Hangzhou, China; Zhijiang Lab, Hangzhou, China*
STEVEN M. DRIEVER • *Centre for Crop System Analysis, Wageningen University and Research, Wageningen, The Netherlands*
PAULINE DUMINIL • *Université Paris-Saclay, CNRS, INRAe, Université Paris Cité, Université d'Evry, Institute of Plant Sciences Paris-Saclay (IPS2), Gif-sur-Yvette, France; Department of Plant Biochemistry, Albrecht-von-Haller Institute for Plant Sciences, University of Gottingen, Gottingen, Germany*
MAIR C. EDWARDS • *Biology Department, Eastern Michigan University, Ypsilanti, MI, USA*
MARION EISENHUT • *Computational Biology, Faculty of Biology, CeBiTec, Bielefeld University, Bielefeld, Germany*
XINYU FU • *Department of Energy-Plant Research Laboratory, Michigan State University, East Lansing, MI, USA*
NATHALIE GLAB • *Université Paris-Saclay, CNRS, INRAe, Université Paris Cité, Université d'Evry, Institute of Plant Sciences Paris-Saclay (IPS2), Gif-sur-Yvette, France*
LUKE M. GREGORY • *Department of Energy-Plant Research Laboratory, Michigan State University, East Lansing, MI, USA; Department of Plant Biology, Michigan State University, East Lansing, MI, USA*
MICHAEL HODGES • *Université Paris-Saclay, CNRS, INRAe, Université Paris Cité, Université d'Evry, Institute of Plant Sciences Paris-Saclay (IPS2), Gif-sur-Yvette, France*
JIANPING HU • *Michigan State University-Department of Energy Plant Research Laboratory and Plant Biology Department, Michigan State University, East Lansing, MI, USA*
KATHRIN JAHNKE • *Plant Physiology Department, University of Rostock, Rostock, Germany*
MARA JAWORSKI • *Biology Department, Eastern Michigan University, Ypsilanti, MI, USA*
HANGJIN JIANG • *Center for Data Science, Zhejiang University, Hangzhou, China*
AUDREY JOHNSON • *Department of Energy – Plant Research Laboratory, Michigan State University, East Lansing, MI, USA*

MATHIEU JOSSIER • *Université Paris-Saclay, CNRS, INRAe, Université Paris Cité, Université d'Evry, Institute of Plant Sciences Paris-Saclay (IPS2), Gif-sur-Yvette, France*
JULIE LALANDE • *Institut de Recherche en Horticulture et Semences, Université d'Angers, Beaucouzé, France*
AARON H. LIEPMAN • *Biology Department, Eastern Michigan University, Ypsilanti, MI, USA*
NICOLE LINKA • *Institute of Plant Biochemistry, Cluster of Excellence on Plant Science (CEPLAS), Heinrich Heine University, Düsseldorf, Germany*
CAROLINE MAUVE • *Université Paris-Saclay, CNRS, INRAe, Université Paris Cité, Université d'Evry, Institute of Plant Sciences Paris-Saclay (IPS2), Gif-sur-Yvette, France*
SHIN-ICHI MIYAZAWA • *Department of Forest Molecular Genetics and Biotechnology, Forestry and Forest Products Research Institute, Tsukuba, Japan*
TIM NIES • *Quantitative and Theoretical Biology, Heinrich Heine University Düsseldorf, Düsseldorf, Germany*
CÉLINE OURY • *Université Paris-Saclay, CNRS, INRAe, Université Paris Cité, Université d'Evry, Institute of Plant Sciences Paris-Saclay (IPS2), Gif-sur-Yvette, France*
RONGHUI PAN • *College of Agriculture and Biotechnology & ZJU-Hangzhou Global Scientific and Technological Innovation Center, Zhejiang University, Hangzhou, China; Zhijiang Lab, Hangzhou, China*
ANASTASIJA PLETT • *Institute of Plant Biochemistry, Cluster of Excellence on Plant Science (CEPLAS), Heinrich Heine University, Düsseldorf, Germany*
CITLALXOCHILT RAMIREZ-LOPEZ • *Biology Department, Eastern Michigan University, Ypsilanti, MI, USA*
MARY BETH ROLLINS • *Department of Plant Pathology and Crop Physiology, Louisiana State University, Baton Rouge, LA, USA*
LUDMILA V. ROZE • *Department of Energy – Plant Research Laboratory, Michigan State University, East Lansing, MI, USA*
STEPHANIE C. SCHMIEGE • *Plant Resilience Institute, Michigan State University, East Lansing, MI, USA; Department of Biology, Western University, London, ON, Canada*
JENNIFER SELINSKI • *Department of Plant Cell Biology, Botanical Institute, Christian-Albrechts University Kiel, Kiel, Germany*
THOMAS D. SHARKEY • *Plant Resilience Institute, Michigan State University, East Lansing, MI, USA; MSU-DOE Plant Research Laboratory, Michigan State University, East Lansing, MI, USA; Department of Biochemistry and Molecular Biology, Michigan State University, East Lansing, MI, USA*
NAVEEN SHARMA • *MSU-DOE Plant Research Laboratory, Michigan State University, East Lansing, MI, USA*
PAUL SOUTH • *Department of Biological Sciences, Louisiana State University, Baton Rouge, LA, USA*
GUILLAUME TCHERKEZ • *Institut de Recherche en Horticulture et Semences, Université d'Angers, Beaucouzé, France; Research School of Biology, Australian National University, Canberra, ACT, Australia*
MAURICIO TEJERA-NIEVES • *Department of Energy-Plant Research Laboratory, Michigan State University, East Lansing, MI, USA; Great Lakes Bioenergy Research Center, Michigan State University, East Lansing, MI, USA*
STEFAN TIMM • *Plant Physiology Department, University of Rostock, Rostock, Germany*

MARVIN VAN AALST • *Quantitative and Theoretical Biology, Heinrich Heine University Düsseldorf, Düsseldorf, Germany*
BERKLEY J. WALKER • *Department of Energy – Plant Research Laboratory, Michigan State University, East Lansing, MI, USA; Department of Plant Biology, Michigan State University, East Lansing, MI, USA; Great Lakes Bioenergy Research Center, Michigan State University, East Lansing, MI, USA*
PHILIPP WESTHOFF • *Institute of Plant Biochemistry, Cluster of Excellence on Plant Science (CEPLAS), Heinrich Heine University, Düsseldorf, Germany*
DONAT WULF • *Computational Biology, Faculty of Biology, CeBiTec, Bielefeld University, Bielefeld, Germany*
YUAN XU • *Department of Energy-Plant Research Laboratory, Michigan State University, East Lansing, MI, USA*

Part I

Assays of Photorespiratory Enzymes

Chapter 1

High Throughput Phosphoglycolate Phosphatase Activity Assay Using Crude Leaf Extract and Recombinant Enzyme to Determine Kinetic Parameters K_m and V_{max} Using a Microplate Reader

Ludmila V. Roze, Audrey Johnson, Luke M. Gregory, Mauricio Tejera-Nieves, and Berkley J. Walker

Abstract

Determining enzyme activities involved in photorespiration, either in a crude plant tissue extract or in a preparation of a recombinant enzyme, is time-consuming, especially when large number of samples need to be processed. This chapter presents a phosphoglycolate phosphatase (PGLP) activity assay that is adapted for use in a 96-well microplate format. The microplate format for the assay requires fewer enzymes and reagents and allows rapid and less expensive measurement of PGLP enzyme activity. The small volume of reaction mix in a 96-well microplate format enables the determination of PGLP enzyme activity for screening many plant samples, multiple enzyme activities using the same protein extract, and/or identifying kinetic parameters for a recombinant enzyme. To assist in preparing assay reagents, we also present an R Shiny buffer preparation app for PGLP and other photorespiratory enzyme activities and a K_m and V_{max} calculation app.

Key words Photorespiration, Phosphoglycolate phosphatase, Recombinant phosphoglycolate phosphatase, Buffer preparation app, Michaelis, Menten, K_m and V_{max} calculation app

1 Introduction

Rubisco-catalyzed oxygenation of ribulose-1,5-bisphosphate (RuBP) initiates the photorespiratory pathway (PR) with the formation of 2-phosphoglycolate (2-PG), which is inhibitory to C3 cycle enzymes and hence net CO_2 assimilation. Phosphoglycolate phosphatase (EC 3.1.3.18) (PGLP) is the "first line of defense" to protect the cell from inhibitory 2-PG by hydrolyzing 2-PG into glycolate and phosphate. This first line of defense is critical due to the scale of rubisco oxygenation, for instance, in all photosynthe-

Berkley J. Walker (ed.), *Photorespiration: Methods and Protocols*, Methods in Molecular Biology, vol. 2792,
https://doi.org/10.1007/978-1-0716-3802-6_1,

sizing organisms on earth rubisco is involved in sequestering ~100 billion tons of oxygen compared to fixation of ~250 billion tons of carbon dioxide annually [1].

During the multistep process of PR, which involves ~10 core enzymes located in chloroplasts, peroxisomes, and mitochondria, 2-PG is processed and its carbon returns to the C3 cycle in the form of 3-phosphoglyceric acid (3-PGA). PGLP catalyzes the first reaction in PR downstream from rubisco and is essential for photosynthetic growth at ambient CO_2 [2, 3]. Furthermore, overexpression of PGLP resulted in enhanced tolerance of *Arabidopsis thaliana* to high light, drought, and high temperature [4]. Moreover, *Rhazya stricta's* particularly high level of photosynthetic and PR activity at elevated temperature is associated with a higher activity of PGLP [5].

Here, we provide a method for measuring PGLP activity in a 96-well format suitable for high-throughput enzyme activity measurements in crude cell extract from leaf tissue and preparation of purified recombinant PGLP, a procedure for obtaining purified recombinant 6xHis-AtPGLP expressed in *Escherichia coli*, and a protocol outlining steps for determining K_m and V_{max} of a purified recombinant AtPGLP using these approaches.

2 Materials

2.1 Plant Material

This approach can be used with any plant material that can be homogenized, but there will likely be some optimization needed based on the activity of the specific samples. For example, we use *Nicotiana tabacum* grown in the greenhouse at ~30 °C until plants form five to seven fully developed leaves. In the protocol described below, we provide specifics for measuring PGLP activity using *N. tabacum* leaves. However, the PGLP activity measurement was successfully accomplished in our laboratory using the outlined protocol in the leaf material from *Rhazya stricta* (Rhazya was grown for ~1.5 years and sampled with six 8.15 mm diameter punches) and birch (at least 1 year old and approximately 120 mg of frozen leaf tissue).

2.2 Total Equipment List

1. Cork borer, ~17 mm diameter.
2. 2 mL screw cap microcentrifuge tubes.
3. 1.5 mL snap cap microcentrifuge tubes.
4. Microplate reader.
5. 96-pin microplate replicator (e.g., Boekel Scientific, cat. # 140500).
6. Refrigerated bench-top centrifuges which work with sample sizes of 0.2–2.0 mL and 15–50 mL.

7. Swinging bucket bench-top centrifuge with adapter for 96-well plate.
8. Pipettes, 2 μL, 20 μL, 300 μL, 1 mL.
9. Multichannel pipette, 300 μL.
10. Tips for pipettes 10 μL, 200 μL, 300 μL, 1 mL.
11. 2 mL glass-to-glass homogenizer (e.g., Kontes Glass Co., Vineland, NJ, USA).
12. Reservoir for multichannel pipet (e.g., Thomas Scientific).
13. Polystyrene 96-well plates.
14. Digital Timer.
15. Parafilm.
16. Oven at 104 °C.
17. Volumetric flask 100 mL.
18. 15 mL and 50 mL plastic tubes.
19. Glass tubes 16 × 125 mm.
20. Autoclavable culture tube caps, plastic, or metal.
21. Sterile toothpicks, pointed.
22. Incubator with orbital shaking platform for 37 °C and 25 °C.
23. Ni-NTA Spin columns.
24. Sonicator.
25. Benchtop liquid nitrogen container.
26. Biosafety cabinet.

2.3 Reagents and Solutions

Use ultrapure sterilized water for all solutions.

2.3.1 PGLP Activity Measurement

1. Extraction buffer (50 mM EPPS buffer, pH 8.0, containing 1 mM EDTA, 10 mM DTT, 0.1% Triton X-100 [v/v], 0.5% polyvinylpyrrolidone, and 10 μL 1× protease inhibitor cocktail, EDTA free). 50 mM EPPS, pH 8.0, and stock solutions 0.5 M EDTA, 1 M $MgCl_2$, and 4 M KCl should be autoclaved and stored at room temperature. Prepare extraction buffer fresh before use. Refer to the buffer calculation app for amounts of chemicals needed to prepare extraction and reaction buffers (Fig. 1).
2. Reaction buffer 50 mM HEPES buffer, pH 7.5, 1 mM EDTA, and 10 mM $MgCl_2$. 50 mM HEPES, pH 7.5, can be prepared in advance, filter-sterilized, and stored at +4 °C. 1 M $MgCl_2$ stock solution should be autoclaved and stored at room temperature. Prepare reaction buffer fresh before use.

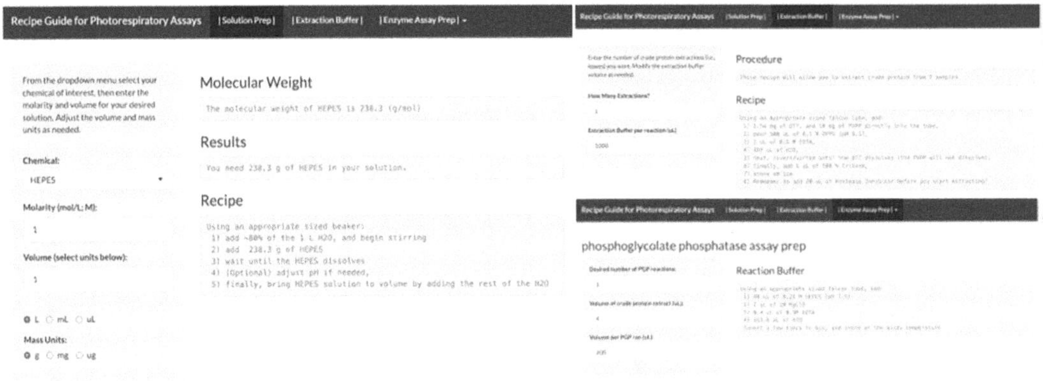

Fig. 1 Custom-written application designed to streamline the solution, extraction buffer, and buffer preparation for photorespiratory enzymes activity assays including PGLP. The custom-built program simplifies the chemical preparation for the PGLP activity assays. Upon executing the app, users can select from three tabs (Solution Prep, Extraction Buffer, Enzyme Assay Prep). The Solution Prep tab automates the recipes for the solutions needed for the PGLP activity assay. User's inputs include chemical of interest, desired concertation, and volume. The Extraction buffer tab automates the volumes needed for the extraction buffer recipes, which scales with the desired number of extractions. The Enzyme Assay Prep tab automates the PGLP buffer recipe using total PGLP reactions, volume of crude protein extract, and total reaction volume as inputs. Download the free custom-built program from GitHub (https:/github.com/L-Gregory/PhotorespiratoryAssay_Recipes)

3. Phosphate stock KH_2PO_4 50 μg/mL. Dehydrate KH_2PO_4 for 1.5–2 h at 104 °C and store in a desiccator before use. The stock can be stored in a tightly closed flask at 4 °C for at least 1 month.
4. 200 mM 2-phosphoglycolic acid lithium salt (2-PG). Aliquot by 0.1 mL, store at −80 °C.
5. Reagent for phosphate determination: 2.5 N sulfuric acid, 0.2 mM antimony potassium tartrate, 4.9 mM ammonium molybdate tetrahydrate, and 30.3 mM ascorbic acid. Prepare the following stocks: the 5 N sulfuric acid stock can be stored at room temperature indefinitely, 4 mM antimony potassium tartrate trihydrate and 32.7 mM ammonium molybdate tetrahydrate can be stored at room temperature for up to a month, and the 100 mM ascorbic acid should be prepared fresh. To make 10 mL of the reagent, mix the solutions before application in the following quantity: 5 mL 5 N sulfuric acid, 0.5 mL 4 mM antimony potassium tartrate, 1.5 mL 32.67 mM ammonium molybdate tetrahydrate, and 3 mL 100 mM ascorbic acid.

2.3.2 Expression and Purification of Recombinant AtPGLP

1. Kanamycin 50 mg/mL stock, aliquot and store at −20 °C.
2. Sterile LB medium, liquid.
3. Sterile LB containing 15 g/L agar and 50 μg/mL kanamycin in Petri dishes.
4. 80% sterile glycerol.

5. 1 M IPTG (isopropyl ß-D-1-thiogalactopyranoside), filter through a sterile 0.22 μm filter using a syringe, aliquot, store at −20 °C.
6. 50 mM Tris–HCl, 300 mM NaCl, pH 8.0. You may prepare the buffer a day before and store at +4 °C.
7. Lysis buffer: 50 mM Tris–HCl, pH 8.0, 300 mM NaCl, 1 mM EDTA, 0.5 mg/mL lysozyme, 0.1 mg/mL DNAse1, 10 mM $CaCl_2$, proteinase inhibitors, 10 mM imidazole, 5% glycerol. Prepare lysis buffer fresh before use.
8. Buffers containing imidazole for binding, washing, and elution are prepared fresh, add imidazole to buffers just before use, keep on ice. Binding buffer 50 mM Tris–HCl, pH 8.0, 300 mM NaCl, and 10 mM imidazole. Washing buffer 50 mM Tris–HCl, pH 8.0, 300 mM NaCl, 20 mM imidazole. Elution buffer 50 mM Tris–HCl, pH 8.0, 300 mM NaCl, and 500 mM imidazole.

2.3.3 Determining Protein Concentration

1. Bradford 1× Dye Reagent, store at +4 °C. Bring to room temperature an aliquot necessary for the construction of the standard curve and processing the number of leaf extracts for a particular experiment.
2. Bovine serum albumin (BSA) dilution series of standards (2000 μg/mL, 1500 μg/mL, 1000 μg/mL, 750 μg/mL, 500 μg/mL, 250 μg/mL, 125 μg/mL), store at −80 °C.

3 Methods

3.1 Preparing Crude Protein Extract

1. For one sample, collect three *N. tabacum* leaf punches each 17 mm in diameter (total area = 6.81 cm^2, approximately 120 mg) (may be more or less depending on species as discussed in Materials section) from fully developed uppermost leaves located next to each other. Using clean forceps quickly place leaf punches into a labeled 2 mL screw cap plastic tube, freeze immediately in liquid N_2, and store at −80 °C. Do not allow to thaw until analysis.
2. Transfer the frozen leaf punches into a glass homogenizer kept on ice, add 1 mL cold extraction buffer, and homogenize manually for approx. 3–5 min until green is removed from the tissues, indicating cells have been ruptured.
3. Transfer the homogenate into a 1.5 mL plastic tube and clarify by centrifugation for 10 min at 13,500 ×*g*, +4 °C. The supernatant represents the crude protein extract and is used in the enzyme assay immediately.

4 Expression and Purification of a Recombinant *Arabidopsis thaliana* PGLP (AtPGLP)

Expression of recombinant N-terminal 6xHis tagged AtPGLP containing no targeting peptide is performed using NEB T7 express *E. coli* cell line transformed with pET28a(+)AtPGLP. The plasmid is available upon request. Under described conditions at least a part of the recombinant protein is soluble according to the SDS-PAGE data. The procedure outlined below results in 90–95% and higher purity of isolated recombinant protein.

4.1 Expression

1. Inoculate 3 mL of sterile B containing kanamycin (50 μg/mL) in a glass tube (16 × 125 mm) with a single colony from agar plate (LB + Kan) using a sterile toothpick or sterile 10 μL tip; perform these procedures in a biosafety hood. Cover the tube with a culture tube cap. Grow the culture at 37 °C overnight at 225 rpm.
2. The next day, inoculate 50 mL in 250 ml flask of pre-warmed sterile LB media containing kanamycin (50 μg/mL), with 0.5 mL of the overnight cultures and grow at 37 °C with vigorous shaking until reaching an OD_{600} of 0.6–0.9 (~2.5 h).
3. Take 1 mL sample in 1.5 mL plastic tube immediately before induction; this sample is to check expression in the non-induced culture. Pellet cells for 10 min at 3500 ×*g* and +4 °C, discard supernatant, freeze the cell pellet on dry ice, and store at −20 °C until analysis by SDS-PAGE.
4. Pre-cool cultures to 25 °C (~5–10 min). Add IPTG to a final concentration of 0.1 mM (50 μL 1 M IPTG/50 mL culture) to induce expression
5. Incubate the cultures overnight at 25 °C until the OD reaches ~2. Collect a second 1 mL sample for the induced culture and pellet cells for 10 min at 3500 ×*g* and +4 °C, then freeze in liquid N_2 and store at −20 °C until analysis by SDS-PAGE.
6. Harvest the cells from the 50 mL culture by centrifugation at 3500 ×*g* for 20 min and +4 °C in 50 mL plastic centrifuge tube.
7. Completely remove supernatant and use the pelleted cells for purification immediately or freeze the cells in liquid N_2 and store cell pellet at −80 °C.

4.2 Purification

Binding, washing, and elution are basically performed according to the manufacturer's Handbook [6] with modifications outlined below. Carry out all steps at +4 °C. Perform purification procedure using 50 mM Tris–HCl, pH 8.0, as K-phosphate buffer, recommended by most manufacturers, interferes with the downstream detection of phosphate used for PGLP activity measurement.

1. Prepare a Ni-NTA column as recommended in the manufacturer's Handbook with modifications: apply 600 μL of 50 mM Tris–HCl, pH 8.0, 300 mM NaCl, and 10 mM imidazole to the column and centrifuge the column at 890 ×*g* for 2 min and +4 °C.
2. Thaw the cell pellet for approx.15 min on ice. Add 1.5 mL lysis buffer to cells obtained from 50 mL induced culture. Mix thoroughly the bacterial pellet and the lysis buffer (*see* **Note 1**).
3. Incubate the combined cell suspension for 1 h on ice with gentle shaking.
4. Keeping the tube with the cell suspension on ice, sonicate 5 × 10 s using a sonicator at high frequency with 10 s intervals for cooling.
5. Centrifuge 20 min, 19,500× *g*, +4 °C (centrifugation could be performed in 1.5–2 mL plastic tubes).
6. Collect the supernatant (this is crude soluble protein extract) into 15 mL conical tube and keep on ice.
7. Apply 600 μL of the supernatant onto a column placed into a 2-mL collection tube and centrifuge the column with the tube at 50 ×*g* for 2 min at +4 °C. The low speed provides time for the enzyme to bind the Ni-matrix. Discard the flow-through but collect 50 μL and store at −20 °C until analysis by SDS-PAGE.
8. Repeat **step** 7 until all supernatant has been loaded.
9. Wash 3 times by applying 600 μL 50 mM Tris–HCl, pH 8.0, 300 mM NaCl, and 20 mM imidazole followed by centrifugation at 890 ×*g* for 2 min at +4 °C. Discard washes but collect 50 μL of washes and store at −20 °C until analysis by SDS-PAGE.
10. Elute the enzyme by applying 300 μL 50 mM Tris–HCl, pH 8.0, 300 mM NaCl, and 500 mM imidazole to the column followed by centrifugation at 890 ×*g* for 2 min at +4 °C.
11. Repeat **step 10**.
12. Pool the eluates (you may keep the eluates separately if there are differences in purity) and dialyze twice against 250 mL 50 mM Tris–HCl, pH 8.0, 0.3 M NaCl at +4 °C, then overnight against 1000 mL 50 mM Tris–HCl, pH 8.0, 0.3 M NaCl, 5% glycerol at +4 °C (*see* **Note 2**).
13. The next day, if protein precipitation is present, remove the precipitated protein by centrifugation at 19,500 ×*g* and +4 °C for 10 min.
14. Measure protein concentration in the supernatant using NanoDrop and/or Bradford protocol for microplate (see below).

15. Add 80% sterile glycerol to result in a final 10% concentration to the dialyzed protein preparation, aliquot, and store at −80 °C.
16. Using SDS-PAGE (not provided in this Chapter) analyze the non-induced and induced cultures, crude soluble extract, flow-through, washes, and eluates.

5 Phosphoglycolate Phosphatase (PGLP) Activity

Methods for PGLP assay in different photosynthesizing species reported in the literature are based on detection of liberated inorganic phosphate during the enzymatic reaction. However, concentrations of the substrate 2-PG, $MgCl_2$, pH, temperature, the buffer used and the assays for phosphate detection were different [7–11].

Here, we describe an assay to measure PGLP activity based on publication by Schwarte and Bauwe [12], with modifications. The pH of PGLP reaction is selected at pH 7.5 based on a report by J. Shen [13] that estimates pH in the plastid stroma at 7.2 ± 0.3. Formation of inorganic phosphate is determined by a procedure published previously [14].

1. Before the assay, dilute the phosphate stock 10 times in 50 mM HEPES buffer, pH 7.5, to obtain a diluted stock 5 μg P/mL. Pipette the diluted KH_2PO_4 stock solution to produce 0.00, 0.05, 0.1, 0.15, 0.20, 025, 0.30, and 0.35 μg P/mL per well in triplicate with the final volume of 200 μL (Table 1). These samples are used for the standard curve construction (Fig. 2).

Table 1
Dilutions of phosphate stock for standard curve

μg P in 200 μL	P stock (5 μg/mL) (μL)	50 mM HEPES, pH 7.5 (μL)
0	0	200
0.05	10	190
0.10	20	180
0.15	30	170
0.20	40	160
0.25	50	150
0.30	60	140
0.35	70	130

To generate standard curve, dilute 50 μg/mL phosphate stock prepared as described in PGLP activity measurement section to obtain 5 μg/mL. Per each well aliquot diluted phosphate stock and 50 mM HEPES, pH 7.5, buffer in volumes indicated in the table. Three technical replicates for each concentration is recommended

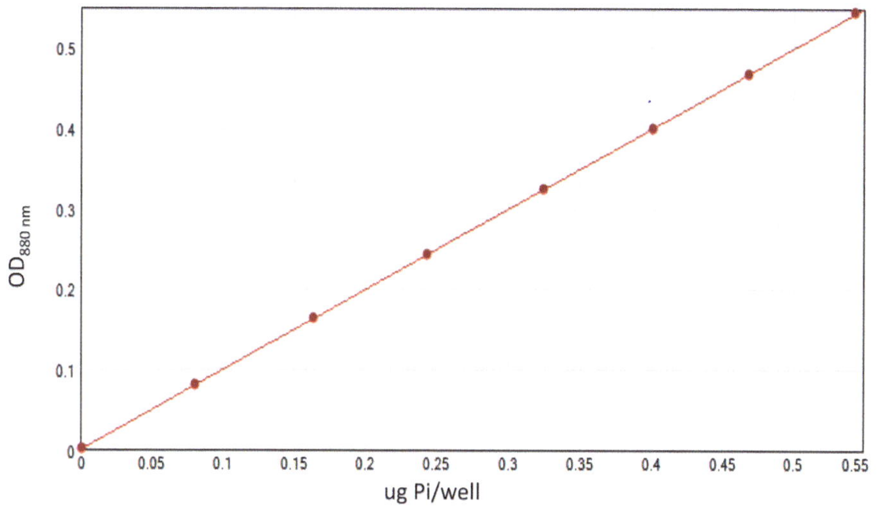

Fig. 2 Standard curve for inorganic phosphate determination with inorganic phosphate concentration on the y-axis and OD_{880} on the y-axis

2. Pipette 194 μL of reaction buffer (50 mM HEPES, pH 7.5, 1 mM EDTA,10 mM $MgCl_2$) into empty wells in triplicate using a 300 μL multichannel pipette and add 4 μL of crude protein extract into each of these wells.
3. Make blank/background well: 200 μL 50 mM HEPES, pH 7.5.
4. In case of recombinant enzyme, instead of crude protein extract, add 4 μL of diluted purified enzyme containing 10 μg of protein.
5. Initiate the enzymatic reaction by adding 2 μL of 200 mM 2-PG using the 96-pin replicator and mix thoroughly. 2 μL of the substrate can be added on the top or the side of the pins of the replicator to accomplish the substrate addition as simultaneously as possible.
6. Incubate the reaction mix for 5 min at 25 °C.
7. Terminate the reaction by adding 32 μL of reagent using a 100 μL multichannel pipette. Add the reagent to the blank/ background well.
8. Cover the plate with the parafilm, keep the plate at room temperature, and take absorbance readings after 45 min at 880 nm using a microplate spectrophotometer (*see* **Notes 3**, **4**, and **5**).
9. Make a control to adjust for phosphate that is not produced during the enzymatic reaction. For the control, combine in the following order: 194 μL reaction buffer, 4 μL crude protein

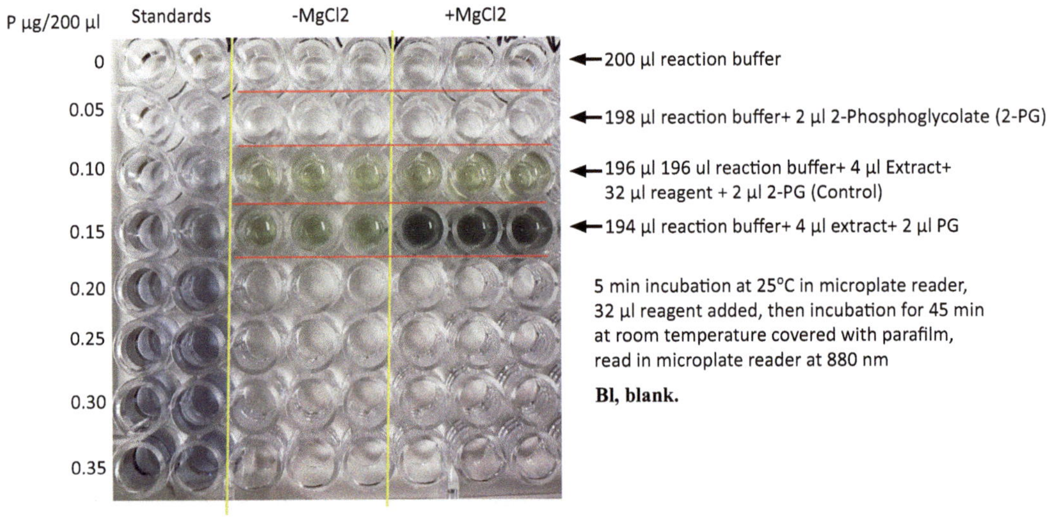

Fig. 3 Layout of phosphate standards and samples with the reaction mixes, controls, and the blank

extract, incubate for 5 min at 25 °C, then add 32 µL of reagent to stop any enzymatic reactions followed by addition of 2 µL of 200 mM 2-PG (Fig. 3).

10. Determine phosphate amounts using the spectrometer's software to plot standard curve and calculate amounts of phosphate produced in the wells containing enzymatic reaction mixes and the control wells. Plate reader software will automatically subtract the blank/background OD value from OD values in all read wells. Subtract the control values from the corresponding values determined in the wells with enzymatic reaction mixes.

6 Protein Concentration

Protein concentration is determined by the Bradford method [15] using a microplate reader. The assay is based on binding Coomassie Brilliant Blue G-250 dye, a component of Bradford 1× Dye Reagent to protein and formation of the blue color product under acidic conditions with the absorbance maximum at 595 nm. The assay is performed in a standard 250 µL microplate assay format using BSA in the linear range of 125–1000 µg/mL, as a standard.

1. Bring to room temperature Bradford 1× Dye Reagent and thoroughly mix by inverting the bottle with the Reagent.
2. Bring BSA standards (2000 µg/mL, 1500 µg/mL, 1000 µg/mL, 750 µg/mL, 500 µg/mL, 250 µg/mL, 125 µg/mL, 0 µg/mL) stored at −80 °C, to room temperature.

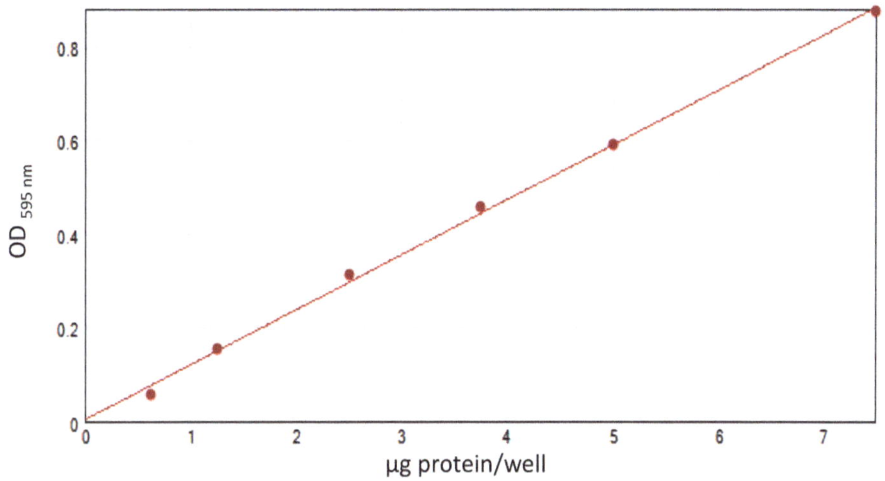

Fig. 4 Standard curve for protein determination with protein concentration on the y-axis and OD_{595} on the y-axis

3. Pipet 5 µL in triplicate of each standard to create 0.00, 0.625, 1.25, 2.50, 3.75, 5.00, and 7.50 µg/well.
4. Pipet 5 µL unknown protein solution in triplicate (it is necessary to make several dilutions) into wells of 96-well microplate; include a blank/background well containing 5 µL water.
5. Using a 300 µL multichannel pipette quickly dispense 250 µL of Bradford Dye Reagent into wells, including the blank well, using clean tips for each set of wells, cover the plate with the parafilm and mix thoroughly in a microplate reader.
6. Incubate the plate at room temperature for 10 min (at least 5 min and no longer than 1 h). Remove the parafilm and record the absorbance at 595 nm.
7. Determine protein concentration using the microplate reader software which automatically plots standard curve and calculates protein concentration in the unknown samples (Fig. 4).

7 Calculating PGLP-Specific Activity

PGLP-specific activity is expressed in µmoles of phosphate liberated per minute per milligram of total protein. An example of calculation:

$$\frac{\Delta P \times 1000}{5 \times 31 \times \text{protein}}.$$

ΔP, amount of P in µg produced in 200 µL reaction mix (control values subtracted).

5, time of the reaction in min.

31, atom weight of *P*.

Protein, μg of crude protein extract in 200 μL reaction mix.

8 K_m and V_{max} Determination for the Purified Recombinant AtPGLP

1. Prepare nine 2-PG solutions of different concentrations according to Table 2, including 0 mM/mL and keep at room temperature.
2. Dilute purified recombinant PGLP to make 0.01 ng/5 μL (0.002 μg/μL) using 50 mM HEPES pH 7.5, keep on ice.
3. Dilute the phosphate stock 10 times in 50 mM HEPES buffer, pH 7.5, to obtain a diluted stock 5 μg P/mL. Pipette the diluted KH_2PO_4 stock solution to produce 0.00, 0.05, 0.1, 0.15, 0.20, 025, 0.30, and 0.35 μg P/mL per well in triplicate with final volume of 200 μL (Table 1). These samples are used for the standard curve construction.
4. Make a blank/background well: add to a well 200 μL 50 mM HEPES, pH 7.5.
5. Pipette 187.5 μL of the reaction buffer into empty wells of 96-well microplate in triplicate for each sample and control using a multichannel pipette.

Table 2
Preparation of 2-PG dilution series for K_m and V_{max} determination using 200 mM 2-PG stock

	For 1 reaction mix		For 10 reaction mixes	
2-PG in 200 μL reaction mix (mM)	2-PG (μL)	H_2O (μL)	2-PG (μL)	H_2O (μL)
	Need 200 mM 2-PG		Need 200 mM 2-PG	
4	4	3.5	40	35
2	2	5.5	20	55
1	1	6.5	10	65
	Need 20 mM 2-PG		Need 20 mM 2-PG	
0.5	5	2.5	50	25
0.3	3	4.5	30	45
0.1	1	6.5	10	65
	Need 2 mM 2-PG		Need 2 mM 2-PG	
0.05	5	2.5	50	25
0.01	1	6.5	10	65

7.5 μL 2-PG is added to 200 μL reaction mix (*see* Subheading 8)

6. Add 7.5 μL 2-PG of different concentrations, including 0 mg/mL, into the wells outlined for the activity measurement.
7. Add 5 μL enzyme solution (0.010 μg PGLP) into the wells containing 2-PG to initiate reaction and mix. Use the 96-pin replicator to add the enzyme.
8. After 5 min stop reaction by adding 32 μL of reagent to each well with the reaction mix.
9. For the control in triplicate, combine in the following order: 187.5 μL reaction buffer, 5 μL enzyme, after 5 min add the reagent to stop any enzymatic reactions followed by the addition of 7.5 μL of 2-PG. The control adjusts for phosphate that is not produced during the enzymatic reaction.
10. Immediately add 32 μL reagent to each well containing phosphate standards and to the blank well.
11. Cover the plate with parafilm.
12. Record OD at 880 nm after 45 min.
13. Determine phosphate amounts using the spectrometer's software to plot standard curves and calculate amounts of phosphate produced in the wells containing enzymatic reaction mixes. Subtract the values obtained in the corresponding control wells.
14. Calculate specific PGLP activity at each concentration of 2-PG as μmoles phosphate/min/mg protein.
15. Determine *Km* and V_{max} using the app (Fig. 5) or other available tools.

9 Notes

1. When larger quantities of purified enzyme are required, it is possible to use 3–4 50 mL separate cultures, or simply one culture of larger volume. Careful attention should be paid to culture aeration. For easier handling, the harvested and frozen cells could be combined for lysis and following sonication in one 50 ml tube.
2. 0.3 M NaCl and 5% glycerol present in Tris buffer help prevent protein precipitation during dialysis.
3. Development of blue color in phosphate assay occurs over time reaching maximum at around 45–60 min.
4. In the event of high phosphate background in phosphate determination assay, all dishes used can be soaked in 3.7% HCl overnight and rinsed thoroughly to remove phosphate contamination.

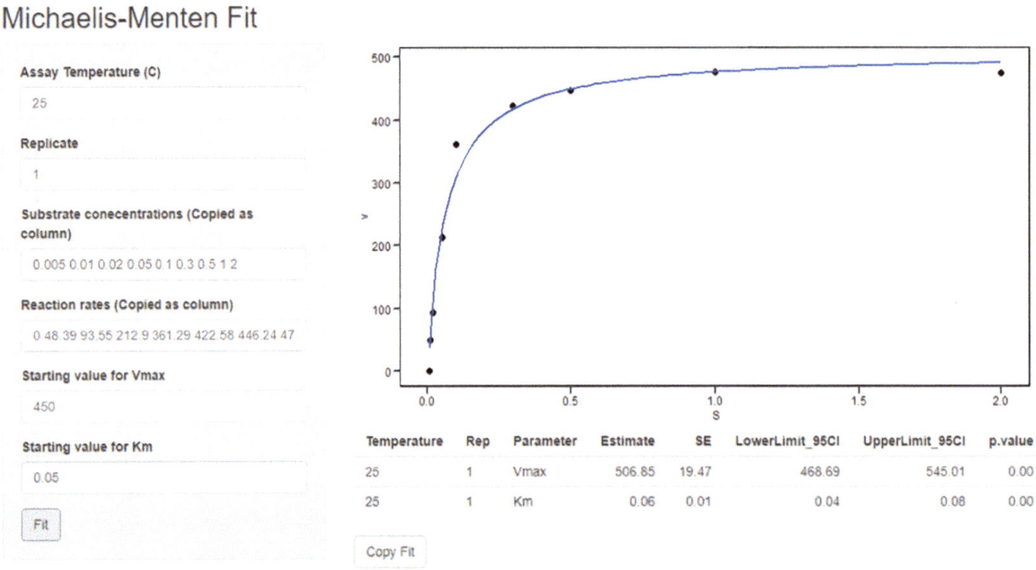

Fig. 5 Custom-written application designed to determine enzyme kinetic parameters the Michaelis constant K_m and the maximum reaction rate of reaction V_{max}. The program fits the activity values (v) and the corresponding substrate concentrations (S) to the non-linear Michaelis-Menten equation $v = (V_{max} * S)/(K_m + S)$. The URL address for using the app is https:/perennialdr.shinyapps.io/MichaelisMentenFit/, the code is available at https:/github.com/PerennialDr/MichaelisMenten_fit. The app requires minimal input from the user as concentrations of 2-PG in mM, the reaction rates as mgP/min or mgP/min/mg protein, and the approximate starting values for K_m and V_{max}

5. Formation of bubbles during phosphate assay interferes with the linearity of the standard curve. Avoid bubbles by gently adding the reagent to the wells and mixing. If bubbles have formed, spinning the plates for 10–15 min at 1500 ×*g* in the swing bucket centrifuge may help.

Acknowledgments

This work was supported by the US Department of Energy Office of Science, Basic Energy Sciences under Award DE- FG02-91ER20021 and the National Science Foundation under grant number 2030337. The work conducted by the U.S. Department of Energy Joint Genome Institute, a DOE Office of Science User Facility, is supported under Contract No. DE-AC02-05CH11231.

References

1. Bathellier C, Yu L-J, Farquhar GD, Coote ML, Lorimer GH, Tcherkez G (2020) Ribulose 1,5-bisphosphate carboxylase/oxygenase activates O2 by electron transfer. Proc Natl Acad Sci USA 117:24234–24242. https://doi.org/10.1073/pnas.2008824117
2. Somerville CR, Ogren WL (1979) A phosphoglycolate phosphatase-deficient mutant of

Arabidopsis. Nature 280:833–836. https://doi.org/10.1038/280833a0

3. Suzuki K, Marek LF, Spalding MH (1990) A photorespiratory mutant of *Chlamydomonas reinhardtii*. Plant Physiol 93:231–237. https://doi.org/10.1104/pp.93.1.231
4. Timm S, Woitschach F, Heise C, Hagemann M, Bauwe H (2019) Faster removal of 2-phosphoglycolate through photorespiration improves abiotic stress tolerance of Arabidopsis. Plants (Basel) 8:563. https://doi.org/10.3390/plants8120563
5. Gregory LM, Roze LV, Walker BJ (2023) Increased activity of core photorespiratory enzymes and CO_2 transfer conductances are associated with higher and more optimal photosynthetic rates under elevated temperatures in the extremophile Rhazya stricta. Plant Cell Environ 46(12):3704–3720
6. Ni-NTA Spin Kit Handbook – QIAGEN. http://www.qiagen.com>resources>download.PDF
7. Christeller JT, Tolbert NE (1978) Phosphoglycolate phosphatase. Purification and properties. J Biol Chem 253:1780–1785
8. Hardy P, Baldy P (1986) Corn phosphoglycolate phosphatase: purification and properties. Planta 168:245–252. https://doi.org/10.1007/BF00402970
9. Husic HD, Tolbert NE (1984) Anion and divalent cation activation of phosphoglycolate phosphatase from leaves. Arch Biochem Biophys 229:64–72. https://doi.org/10.1016/0003-9861(84)90130-9
10. Belanger FC, Ogren WL (1987) Phosphoglycolate phosphatase: purification and preparation of antibodies. Photosynth Res 14:3–13. https://doi.org/10.1007/BF00019588
11. Norman EG, Colman B (1991) Purification and characterization of phosphoglycolate phosphatase from the cyanobacterium *Coccochloris peniocystis1*. Plant Physiol 95:693–698
12. Schwarte S, Bauwe H (2007) Identification of the photorespiratory 2-Phosphoglycolate phosphatase, PGLP1, in Arabidopsis. Plant Physiol 144:1580–1586. https://doi.org/10.1104/pp.107.099192
13. Shen J, Zeng Y, Zhuang X, Sun L, Yao X, Pimpl P, Jiang L (2013) Organelle pH in the Arabidopsis endomembrane system. Mol Plant 6:1419–1437. https://doi.org/10.1093/mp/sst079
14. Su-Cheng P, Chung-Cheng Y, Riley JP (1990) Effects of acidity and molybdate concentration on the kinetics of the formation of the phosphoantimonylmolybdenum blue complex. Anal Chim Acta 229:115–120. https://doi.org/10.1016/S0003-2670(00)85116-8
15. Quick Start™Bradford Protein Assay. https://www.bio-rad.com/webroot/web/pdf/lsr/literature/4110065A.pdf

Chapter 2

Measuring Phosphoglycolate Phosphatase Activity Using NMR Detection of Glycolate

Caroline Mauve, Céline Oury, Nathalie Glab, and Michael Hodges

Abstract

Besides the historical and traditional use of nuclear magnetic resonance (NMR) spectroscopy as a structure elucidation tool for proteins and metabolites, its quantification ability allows the determination of metabolite amounts and therefore enzymatic activity measurements. For this purpose, ^{1}H-NMR with adapted water pulse pre-saturation sequences and calibration curves with commercial standard solutions can be used to quantify the photorespiratory cycle intermediates, 2-phosphoglycolate and glycolate, associated with the phosphoglycolate phosphatase reaction. The intensity of the ^{1}H-NMR signal of glycolate produced by the activity of purified recombinant *Arabidopsis thaliana* PGLP1 can therefore be used to determine PGLP1 enzymatic activities and kinetic parameters.

Key words NMR, 2-Phosphoglycolate phosphatase, Photorespiration, Enzyme activity, Glycolate, 2-Phosphoglycolate

1 Introduction

Phosphoglycolate phosphatase (PGLP; EC3.1.3.18) belongs to the haloalkanoic acid dehalogenase protein superfamily [1]. In plants, this enzyme carries out the dephosphorylation of 2-phosphoglycolate (2-PG) to give glycolate. It is an important photorespiratory cycle enzyme as it removes toxic 2-PG produced by the oxygenase activity of Rubisco [2]. Several techniques to measure PGLP activity can be found in the literature including PGLP-derived phosphate release measured by colorimetric assays after the formation of phosphomolybdate [3–6], a radiometric assay using [^{32}P]-labeled 2-PG [7] and more recently a coupled enzymatic reaction using recombinant glycolate oxidase (GOX), *o*-dianisidine and horseradish peroxidase that allows quantification of PGLP-made glycolate via GOX-produced H_2O_2 by spectrophotometry at 440 nm ([8–10] and also *see* Chap. 3 in this book). To overcome problems related to phosphate contaminations and the

Berkley J. Walker (ed.), *Photorespiration: Methods and Protocols*, Methods in Molecular Biology, vol. 2792,
https://doi.org/10.1007/978-1-0716-3802-6_2,

interference of enzymatic assays in the presence of oxidants like H_2O_2 that interfere with the GOX-coupled-reaction assay, a nuclear magnetic resonance (NMR) protocol was therefore developed to measure PGLP activity.

NMR is an analytical chemistry technique based on the fact that many nuclei have a spin (e.g., +1/2 or −1/2 for ^1H), and in the absence of an external magnetic field, these nuclear spins are randomly oriented. When an external magnetic field is applied (B0; in this protocol B0 is 400 MHz) to these nuclei, two possible orientations are possible, either parallel or antiparallel to the field; however, these two positions do not have the same energy. When an oscillating magnetic field pulse (B1) is applied perpendicular to B0, the spins are moved from a low-energy position to a high-energy position. At the end of the pulse, the spins recover their initial orientation with a specific frequency (relaxation phenomenon) that depends on their initial spin orientation and the energy acquired with the B1 pulse. This radio-frequency-relaxation signal can produce an interferogram called the Free Induction Decay (FID), which is converted into a spectrum (chemical shift in ppm) via Fourier transformation. The chemical shift relies on the nucleus environment specific to each molecule and it is measured with respect to a reference compound. The result is a spectrum with the chemical shift on the x-axis and the signal intensity on the y-axis. Each ^1H of the molecule can have a signal in the B0 field. Processed peak areas are then quantified using calibration curves made using highly pure metabolite solutions.

Metabolite quantifications by NMR are now widely used in various domains such as metabolomics [11] and flux analyses [12]. In the case of complex mixtures (i.e., biological samples such as leaf extracts), signal overlaps often occur and lead to a poor resolution and thus a poor quantification. For our purpose, a low magnetic field was used to analyze a purified enzyme with only the addition of substrate and required cations. In such conditions, the resulting NMR spectra showed no overlaps and they were easy to identify. This allowed monitoring of the substrate (2-PG) and the product (glycolate) of the enzymatic reaction of PGLP. For such enzymatic NMR measurements, the use of adapted pulse sequences can drastically reduce the water signal and greatly improve spectra quality to obtain a good signal-to-noise ratio for the peak(s) of interest. Several possible pre-saturation NMR sequences are possible and they are well documented [13, 14]. Here excitation sculpting [15] was used as it gave the best result, suppressing the water signal and generating excellent resolution of both 2-PG and glycolate spectra. This allowed the quantification of PGLP-produced glycolate from its ^1H-NMR spectra and the calculation of PGLP activity.

2 Materials

2.1 PGLP Activity

1. Precision pipettes (0-2 μL, 0–10 μL, 0–20 μL, 0–200 μL, and 0–1000 μL).
2. Precision balance.
3. pH meter.
4. Glass beakers (250 mL, 500 mL).
5. Glass bottles (250 mL).
6. Graduated cylinders (250 mL, 500 mL).
7. Magnetic stirring bar.
8. Magnetic stirrer.
9. Vortex.
10. 1.5 mL Eppendorf SAFE Lock tubes.
11. 50 mL plastic tubes.
12. Dry block heater.
13. Timer.
14. Plastic spatulas.
15. Icebox containing ice.
16. Nitrile gloves.
17. Protective glasses.

2.2 NMR Analyses

1. NMR machine (*see* Fig. 1 and **Note 1**).
2. Analytical software (*see* **Note 2**).
3. Sample tray (*see* **Note 3**).
4. 5 MM, 400 MHZ, 7″ Standard NMR Tubes.

2.3 Reagents, Solutions, and Buffers for the PGLP Reaction and NMR Analysis

1. Purified PGLP enzyme (*see* **Note 4**).
2. Ultrapure water.
3. 12.1 N Hydrochloric acid (*see* **Note 5**).
4. Ultrapure Tris-base; 0.5 M Tris–HCl, pH 8, stock solution (*see* **Note 6**).
5. $MgCl_2.6H_2O$ (purity ≥99.0%); 0.5 M stock solution (*see* **Note 7**).
6. EDTA-free Protease Inhibitor Cocktail Tablets (*see* **Note 8**).
7. 2-Phosphoglycolic acid (Lithium salt, purity ≥95.0); 250 mM stock solution (*see* **Note 9**).
8. Glycolic acid (analytical standard, purity 99%); 0.5 M stock solution (*see* **Note 10**).
9. Deuterium oxide (D_2O) 99.9 atom % D contains 0.75 wt. % 3-(trimethylsilyl)propionic-*2,2,3,3-*d_4 acid, sodium salt.

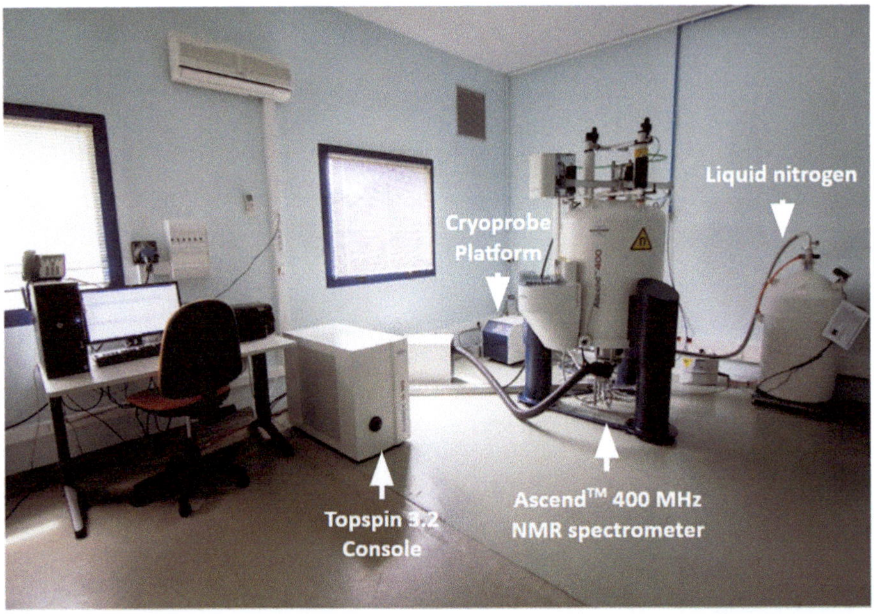

Fig. 1 Overview of the NMR set-up used for measurements in the protocol. The 400 MHz NMR spectrometer is equipped with a cryoprobe platform that requires cooling by liquid nitrogen and a Topspin 3.2 console

3 Methods

3.1 PGLP Reaction

1. Prepare the initial reaction mix: Pipette 5 mL of 0.5 M Tris–HCl, pH 8 and 1 mL of 0.5 M $MgCl_2$ into a 50 mL plastic tube and add ultrapure water to a final volume of 50 mL. Add 1 tablet of antiprotease and mix to dissolve (*see* **Note 8**).
2. Determine the volume of initial reaction mix (*see* **step** 1 in Subheading 3.1 above; *see* **Note 11**), for a final reaction volume of 1 mL that will contain 1 mM 2-PG and 1 μg purified desalted PGLP (for 2-PG, use 40 μL of 25 mM 2-PG and *see* **Note 9**; for purified desalted PGLP *see* **Note 4**).
3. Pipette the required volume of the initial reaction mix into 1.5 mL Eppendorf tubes.
4. Place the Eppendorf tubes at 30 °C in a dry block heater and allow them to equilibrate for at least 3 min.
5. Add 40 μL of 25 mM 2-PG (1 mM final concentration), mix well with a disposable spatula, and leave the spatula in the open tube.
6. Start the PGLP reaction by adding 1 μg of purified desalted PGLP (*see* **Note 4**) to the reaction mix containing 2-PG, rapidly mix with the plastic spatula, and leave the spatula in the open tube. At the same time, start the timer then stop the reaction after 20 s (*see* **Note 12**) by adding 60 μL of 12.1 N hydrochloric acid (*see* **Note 5**), rapidly mix with the plastic

spatula, and remove it. Carefully close the tube, vortex and keep on ice until all PGLP reactions are completed. Repeat **steps 4** and **5** in Subheading 3.1 for each reaction to be carried out. Carry out a time zero control by adding 60 μL of 12.1 N hydrochloric acid to the reaction mix containing 2-PG in the absence of PGLP. Rapidly mix with the plastic spatula, remove it, and then add 1 μg of purified desalted PGLP. Carefully close the tube, vortex, and keep on ice.

7. Use 490 μL of the HCl-stopped final reaction mix for NMR detection of produced glycolate and remaining 2-PG. This can be stored at 4 °C (on ice) for a short time before carrying out the automated NMR analysis.

3.2 Identification and Quantification of Glycolate by ^{1}H-NMR to Calculate Purified PGLP Enzymatic Activity

3.2.1 Data Acquisition to Determine PGLP-Produced Glycolate

1. Pipette 490 μL of each final reaction mix (*see* **step 5** or **6** in Subheading 3.1) into NMR tubes.
2. Add 10 μL of D_2O with 0.75 wt. % sodium 3-trimethylsilyl-*2,2,3,3-d*$_4$-propionate (TMSP) as a reference (*see* **Note 13**).
3. Place tubes into the auto-sampler of the NMR machine.
4. Start NMR sequence runs (*see* **Note 14**). We apply Bruker excitation sculpting sequence (zgesgp) with NS 128 scans (*see* **Note 15**). Automatically lock, tune, and shim were applied using the program IconNMR (*see* **Note 16**).
5. Acquire data at 277 K without tube spinning for 9.5 min (the time required for 128 scans).

3.2.2 Data Processing and Analysis

We analyze all data using TopSpin software (Bruker).

1. Process the FID signals by performing exponential line broadening and calculate the Fourier transformation (*see* **Note 17**).
2. Adjust the phase, apply baseline correction, and calibrate with TMSP.
3. Integrate the singlet peak of glycolate (4.21 ppm) and the reference (*see* Fig. 2).
4. Check that the standard peak has a correct variation (*see* **Note 18**).
5. Quantify glycolate using the calibration curve (*see* Fig. 3 and **Note 19**).
6. Calculate the initial velocity of the PGLP reaction (*see* **Note 20**).

3.3 Determination of PGLP Kinetic Parameters

1. Prepare an adequate 2-PG concentration range using the 250 mM stock solution and 50 mM Tris–HCl, pH 8 (*see* **Note 21**).
2. Carry out a PGLP reaction for each 2-PG concentration to be tested by adding the appropriate volume of 2-PG solution to the reaction mix as previously described (*see* Subheading 3.1).

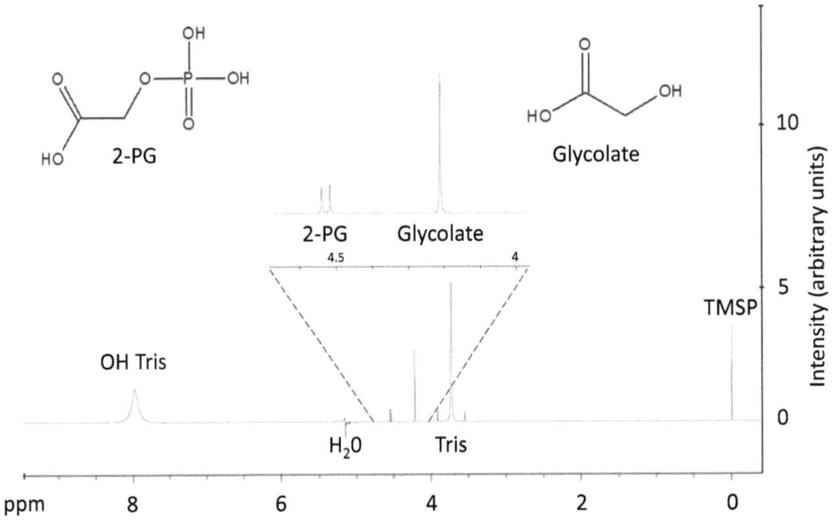

Fig. 2 Attribution of ^{1}H-NMR spectrum peaks. To identify the peaks of interest in the ^{1}H-NMR spectrum, a 1 mL PGLP enzyme assay reaction mix containing 2 mM 2-PG and 2 mM glycolate and 60 μL of 12.1 N HCl was prepared before the addition of 1 μg of recombinant PGLP. The CH_2 of glycolate was identified at 4.21 ppm and since the protons are equivalent, a singlet can be observed. In the case of 2-PG, the protons of the CH_2 group are not symmetric because of the phosphate group, and therefore a doublet can be observed at 4.53 ppm

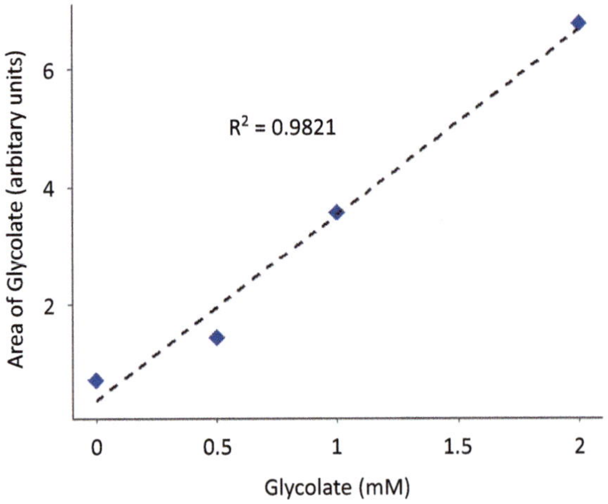

Fig. 3 Glycolate calibration curve. To calculate the quantity of PGLP-produced glycolate, a calibration curve is used where NMR singlet peak area of glycolate is plotted against known glycolate concentrations and a linear regression is carried out

3. For each 2-PG concentration to be tested, quantify by NMR the glycolate produced during the reaction time, and calculate the initial velocity (*see* **Note 20**).
4. Fit against the Michaelis-Menton equation to determine the kinetic parameters (*see* **Note 22**).

4 Notes

1. We used an Ascend 400 MHz NanoBay NMR spectrometer equipped with an AVANCE III HD console and a Prodigy BBO cryoprobe (Bruker).
2. TopSpin software version 3.2 (Bruker) was used for data analyses.
3. SampleCase™ is a 24-place random-access automation system used for automatic NMR data acquisition.
4. We used purified desalted recombinant His-tagged *At*PGLP1 (both H_{10}-*At*PGLP1 and H_6-MBP-*At*PGLP1) for our experiments (*see* Chap. 8). PGLP1 was diluted to 50 ng protein/μL with the appropriate desalting buffer.
5. As hydrochloric acid is an irritant and highly corrosive, it must be handled with caution under a chemical fume hood. Protective glasses and gloves must be worn. It is strongly recommended to keep gloves on throughout the NMR activity procedure.
6. To prepare the 0.5 M Tris–HCl stock solution at pH 8: Weigh 30.29 g of Tris-base and dissolve in 450 mL of ultrapure water in a glass beaker with a magnetic stirrer, adjust the pH to 8 using HCl. While concentrated HCl (12.1 N) can be used (*see* **Note 5**), it is best to adjust the pH with a lower concentrated HCl solution (1 N) when nearing the desired pH. Then adjust the volume to 500 mL in a 500 mL graduated cylinder with ultrapure water, homogenize, and store in 250 mL glass bottles. Tris–HCl stock solution can be stored for several months at 4 °C.
7. To prepare the 0.5 M $MgCl_2$ stock solution: Weigh 20.33 g of $MgCl_2.6H_2O$ and dissolve in 180 mL of ultrapure water in an appropriate glass beaker before adjusting the volume with ultrapure water to 200 mL in a 250 mL graduated cylinder and homogenize and store in a 250 mL glass bottle. $MgCl_2$ stock solution can be stored for several months at 4 °C.
8. One tablet per 50 mL of reaction buffer is added and dissolved either by using a magnetic stirring bar and a magnetic stirrer or by vortexing.
9. To obtain a 250 mM 2-PG stock solution: the entire 50 mg vial of 2-PG is dissolved by adding 1281.8 μL of Tris–HCl 50 mM pH 8 directly into the vial. Aliquot into small volumes (max. 200 μL) in 1.5 mL Eppendorf tubes to avoid freezing/unfreezing cycles. It can be kept at −20 °C for one year. When required, the 25 mM 2-PG working solution is prepared with 50 mM Tris–HCl, pH 8 from the 250 mM 2-PG stock solution, and can be kept on ice (during use) and only frozen once at −20 °C.

10. To prepare the 0.5 M glycolate solution: Weigh 38 mg of glycolic acid and dissolve it in 1 mL of 50 mM Tris–HCl, pH 8. The solution can be kept aliquoted at −20 °C for 1 year.
11. Volumes and anti-protease tablet number (*see* **Note 8**) should be adapted if more or less reaction mix is required.
12. For each PGLP sample, check the linearity of the enzymatic reaction using different protein amounts and times.
13. All NMR spectra were referenced to TMSP at a 0.00 ppm chemical shift. TMSP was used as it is water soluble and thus compatible with aqueous samples.
14. Topspin "iconNMR automatization" can be programmed to automatically analyze 24 samples (placed in the sample tray; *see* **Note 2**).
15. Optimization of scan number should be done to obtain a good signal-to-noise ratio. More scans give a bigger signal.
16. This step can be done manually to correct tune and shim so as to improve spectra quality and thus obtain the best signal-to-noise ratio.
17. We applied an exponential apodization correction to the Fourier transformation.
18. For each series of experiments, the absolute area of TMSP is measured and the standard deviation is calculated, if lower than 10% then all measurements are validated.
19. For each series of experiments, calibration curves are made using the reaction mix without PGLP and different concentrations of 2-PG and glycolate (0, 0.5, 1, and 2 mM). Data acquisition was confirmed to be quantitative when peak areas were proportional to the amount of the molecules of interest (R^2 close to 1).
20. To calculate PGLP activity: Use the calibration curve to calculate the amount of glycolate (in μmol) produced during the time-course of the reaction (here it was 20 s) for the quantity of PGLP tested (here it was 1 μg). Use this information to calculate PGLP activity in μmol glycolate/min/mg protein.
21. Appropriate 2-PG diluted working solutions are prepared with 50 mM Tris–HCl, pH 8. 25 mM 2-PG is prepared from the 250 mM 2-PG stock solution (*see* **Note 9**). 2.5 mM is prepared from the fresh 25 mM 2-PG solution. They are kept on ice (during use) and only frozen once at −20 °C.
22. To determine PGLP kinetic parameters: Measure PGLP activity with an extended 2-PG concentration range (e.g., from 0.1 to 2.5 mM; under 0.1 mM is too low for NMR detection). Use 2-PG working or stock solutions (*see* **Note 21**) so as to pipette between 20 and 100 μL into the Eppendorf tube. Depending on the added volume of 2-PG, adjust the initial reaction mix volume.

Acknowledgments

We would like to thank Pauline Duminil who helped to develop this protocol during her PhD in the Metaboactions team at IPS2. We would like to thank the French Agence Nationale de la Recherche (ANR) for financial support: ANR-10-LABX-0040-SPS, ANR-11-IDEX-0003-02, ANR-14-CE19-0015 REGUL3P. PD received PhD fellowships from the French Education Ministry. We acknowledge the "Plateforme Métabolisme-Métabolome" of IPS2 for the use of their NMR facility.

References

1. Burroughs AM, Allen KN, Dunaway-Mariano-D, Aravind L (2006) Evolutionary genomics of the HAD superfamily: understanding the structural adaptations and catalytic diversity in a superfamily of phosphoesterases and allied enzymes. J Mol Biol 361:1003–1034
2. Flügel F, Timm S, Arrivault S et al (2017) The photorespiratory metabolite 2-phosphoglycolate regulates photosynthesis and starch accumulation in Arabidopsis. Plant Cell 29:2537–2551
3. Richardson KE, Tolbert NE (1961) Phosphoglycolic acid phosphatase. J Biol Chem 236: 1285–1290
4. Husic HD, Tolbert NE (1984) Anion and divalent cation activation of phosphoglycolate phosphatase from leaves. Arch Biochem Biophys 229:64–72
5. Norman EG, Colman B (1991) Purification and characterization of phosphoglycolate phosphatase from the cyanobacterium *Coccochloris peniocystis*. Plant Physiol 95:693–698
6. Bauwe H (2017) Measurement of enzyme activities. In: Fernie AR, Bauwe H, Weber APM (eds) Photorespiration: methods and protocols. Springer, New York, pp 31–50
7. Rose ZB, Grove DS, Seal SN (1986) Mechanism of activation by anions of phosphoglycolate phosphatases from spinach and human red blood cells. J Biol Chem 261:10996–11002
8. Macheroux P, Massey V, Thiele DJ, Volokita M (1991) Expression of spinach glycolate oxidase in *Saccharomyces cerevisiae*: purification and characterization. Biochemistry 30:4612–4619
9. Stenberg K, Lindqvist Y (1996) High-level expression, purification, and crystallization of recombinant spinach glycolate oxidase in *Escherichia coli*. Protein Expr Purif 8:295–298
10. Dellero Y, Mauve C, Boex-Fontvieille E et al (2015) Experimental evidence for a hydride transfer mechanism in plant glycolate oxidase catalysis. J Biol Chem 290:1689–1698
11. Kim HK, Choi YH, Verpoorte R (2010) NMR-based metabolomic analysis of plants. Nat Protoc 5(3):536–549
12. Abadie C, Tcherkez G (2021) ^{13}C Isotope labelling to follow the flux of photorespiratory intermediates. Plants 10:427. https://doi.org/10.3390/plants10030427
13. McKay RT (2009) Recent advances in solvent suppression for solution NMR: a practical reference. Annu Rep NMR Spectr 66:33–76
14. Price WS (1999) Water signal suppression in NMR spectroscopy. Annu Rep NMR Spectr 38:289–354
15. Roumestand C, Canet D (2000) Extending the excitation sculpting concept for selective excitation. J Magn Reson 147(2):331–339

Chapter 3

Determination of Phosphoglycolate Phosphatase Activity via a Coupled Reaction Using Recombinant Glycolate Oxidase

Pauline Duminil, Céline Oury, Michael Hodges, and Nathalie Glab

Abstract

Phosphoglycolate phosphatase (PGLP) dephosphorylates 2-phosphoglycolate to glycolate that can be further metabolized to glyoxylate by glycolate oxidase (GOX) via an oxidative reaction that uses O_2 and releases H_2O_2. The oxidation of *o*-dianisidine by H_2O_2 catalyzed by a peroxidase can be followed in real time by an absorbance change at 440 nm. Based on these reactions, a spectrophotometric method for measuring PGLP activity using a coupled reaction with recombinant *Arabidopsis thaliana* GOX is described. This protocol has been used successfully with either purified PGLP or total soluble proteins extracted from Arabidopsis rosette leaves.

Key words Phosphoglycolate phosphatase, Glycolate oxidase, Photorespiration, Enzyme activity, Coupled reaction, Spectrophotometry

1 Introduction

Phosphoglycolate phosphatase (PGLP; EC 3.1.3.18) belongs to the haloacid dehalogenase (HAD) superfamily, a group of hydrolases associated with a diversity of activities including phosphatase, dehalogenase, phosphoamidase, sugar phosphomutase, and phosphonatase activities. Most of the phosphatases in this family target small metabolites while proteins are only targeted in some rare cases [1–3]. Plant PGLP was first reported in 1961, and it catalyzes the dephosphorylation of 2-phosphoglycolate (2-PG) to glycolate with the release of inorganic phosphate (Pi) [4]. Over the years, PGLP enzymes from several organisms have been characterized and several methods have been developed to measure their activity. Many studies used variations of methods aiming to measure the Pi released during 2-PG dephosphorylation. The Pi content of a solution can be determined via a colorimetric assay relying on the reduction of phosphomolybdate molecules detectable at 820 nm

Berkley J. Walker (ed.), *Photorespiration: Methods and Protocols*, Methods in Molecular Biology, vol. 2792,
https://doi.org/10.1007/978-1-0716-3802-6_3,

[4–6]. A recent adaptation of this protocol was described in Ref. [7]. Alternatively, the complex formed between phosphate, molybdate, and malachite green can be detected at 650 nm [8–12]. Another method using 2,7-dihydroxynaphthalene and concentrated H_2SO_4 allows PGLP-produced glycolate to be measured spectrophotometry at 535 nm when the assay mixture is heated to produce a colored solution [4, 13]. A radiometric assay with [^{32}P]-labeled 2-PG was also developed and used to measure PGLP activity from spinach and human red blood cells [9]. However, such methods do not allow for the real-time measurement of PGLP activity and great care must be taken to eliminate phosphate contamination.

The determination of an enzyme activity by a coupled reaction using another enzyme (or a series of other enzymes) to allow real-time spectrophotometric measurements is common practice. In leaf primary metabolism, a striking example is the determination of phosphoenolpyruvate carboxylase activity by following the oxidation of NADH at 340 nm in a coupled reaction with NADH-malate dehydrogenase [14, 15]. Another example is the measurement of 2,3-bisphosphoglycerate-independent phosphoglycerate mutase activity using a complex three-enzyme coupled reaction (with enolase, pyruvate kinase, and lactate dehydrogenase) to follow its activity in real time by measuring NADH oxidation at 340 nm [16, 17]. In leaves, photorespiratory PGLP activity produces glycolate that is further metabolized to glyoxylate by GOX. In *Arabidopsis thaliana,* this is achieved by either GOX1 or GOX2 [18]. We developed a spectrophotometric assay that measures glycolate produced by the activity of PGLP using a coupled-reaction system with recombinant *Arabidopsis thaliana* glycolate oxidase 1 (*At*GOX1). The GOX reaction produces H_2O_2 that can oxidize *o*-dianisidine in the presence of horseradish peroxidase (HRP) and the oxidized *o*-dianisidine is detected by spectrophotometry at 440 nm (*see* Fig. 1) [19–22]. This coupled reaction allows the indirect measurement of PGLP activity in real time. This real-time spectrophotometric method allows a precise determination of the linear initial velocity of the PGLP reaction and immediately indicates whether the measuring conditions are adequate to obtain this important kinetic parameter.

2 Materials

2.1 Frequently Used Materials

1. Precision pipettes (0-2 μL, 0–10 μL, 0–20 μL, 0–200 μL, 0–1000 μL).
2. pH meter.
3. Plastic disposable cuvettes (1 mL).
4. Glass bottles (100 mL, 250 mL, 500 mL).

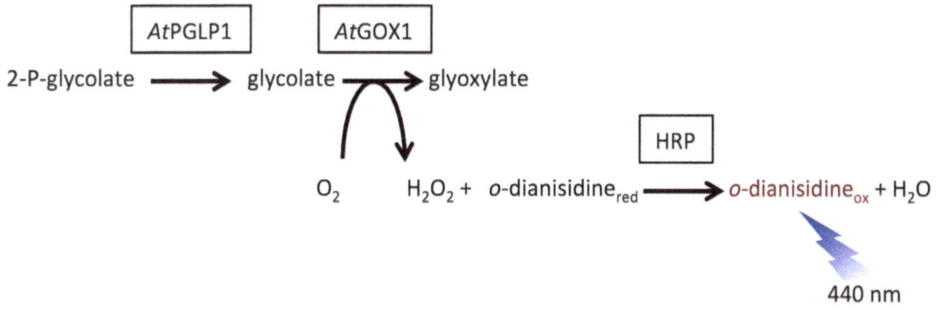

Fig. 1 The coupled reaction to measure PGLP activity using GOX. 2-P-glycolate: 2-phosphoglycolate, AtPGLP1: *Arabidopsis thaliana* phosphoglycolate phosphatase 1, AtGOX1: *Arabidopsis thaliana* glycolate oxidase 1, o-dianisidine$_{red/ox}$: *o*-dianisidine reduced/oxidized, HRP: horseradish peroxidase

5. Graduated cylinders (100 mL, 250 mL, 500 mL).
6. Magnetic stirring bar.
7. Magnetic stirrer.
8. Precision balance.
9. Vortex.
10. Nitrile gloves.
11. Protective face mask.
12. Icebox containing ice.
13. Aluminum foil.

2.2 Reagents

1. Ultrapure water.
2. Tris-Base.
3. $MgCl_2.6H_2O$ (≥99.0% purity).
4. *o*-Dianisidine dihydrochloride (≥95% purity) (*see* **Note 1**).
5. Horse radish peroxidase (HRP).
6. Glycolic acid (99% purity).
7. 2-Phosphoglycolate (Lithium salt ≥95.0% purity).
8. EDTA free Protease Inhibitor Cocktail Tablets (*see* **Note 2**).

2.3 Stock Solutions for Buffers and Reagents

The preparation of stock solutions is detailed in **Note 3** for buffers and in **Notes 4–7** for reagents. The concentrations are as follows:

1. 0.5 M Tris–HCl, pH 8 (*see* **Note 3**).
2. 0.5 M $MgCl_2$ (*see* **Note 3**).
3. 20 mM *o*-dianisidine (*see* **Notes 1** and **4**).
4. 1000 U/mL HRP (*see* **Note 5**).
5. 0.5 M glycolic acid (*see* **Note 6**).
6. 250 mM 2-PG (*see* **Note 7**).

For the duration of an experiment, all reagents are stored on ice.

2.4 Leaf Total Soluble Protein Extraction

1. 300 mg of *A. thaliana* rosette leaves stored at −80 °C until use (*see* **Note 8**).
2. 2 mL Eppendorf tubes.
3. Stainless steel beads (3 mm).
4. Ball mill.
5. Liquid nitrogen.
6. Refrigerated bench centrifuge for 2 mL Eppendorf tubes.
7. 2.5 mL desalting columns pre-packed with G-25 Sephadex beads.
8. Extraction buffer (*see* **Notes 2** and **9**).
9. Spectrophotometer with selectable wavelength for single reads at 595 nm.
10. Commercial Bradford reagent.
11. Bovine serum albumin (BSA) (cold ethanol fraction, pH 5.2, ≥96% storage +4 °C or equivalent).

2.5 PGLP1 Activity Measurement via a Coupled Reaction

1. Spectrophotometer with temperature-controlled automated cuvette changer and integrated analytical software.
2. Reaction buffer (*see* **Notes 2** and **9**).
3. Purified recombinant *At*GOX1 protein (*see* **Note 10**).
4. Leaf total soluble protein (50 μg) or purified PGLP protein (5 μg) (*see* **Note 11**).
5. Disposable plastic stirring spatula.

3 Methods

3.1 Leaf Total Soluble Protein Extraction

1. Collect approximately 300 mg of *A. thaliana* rosette leaves (*see* **Note 8**) in 2 mL Eppendorf tubes containing two stainless steel beads.
2. Immediately freeze the samples in liquid nitrogen and keep them at −80 °C until use (*see* **Note 12**).
3. Grind the harvested frozen leaf sample using a ball mill for 60 s (2 × 30 s) at 30 Hz and re-suspend the resulting powder in 400 μL of extraction buffer (*see* **Notes 9** and **13**).
4. Centrifuge the sample for 15 min at 22000× *g* using a fixed angle rotor benchtop centrifuge at 4 °C and transfer the supernatant to a clean 2 mL Eppendorf tube. Avoid the pellet.
5. Repeat the centrifugation step to clear the lysate from all leaf debris.
6. Transfer the supernatant to a clean 2 mL Eppendorf. Keep on ice before carrying out the desalting step.

3.2 Desalting of Leaf Soluble Protein Extracts and Protein Concentration Determination

1. Equilibrate the gel filtration chromatography column by adding extraction buffer to the top of column and by letting gravity flow through the resin. Repeat this three times. Approximately 10 mL of buffer will be used.
2. When no more extraction buffer elutes from the column, slowly load the leaf total soluble protein extract (drop by drop) onto the resin surface using a precision pipette.
3. After elution stops add 1 mL of extraction buffer to the surface of the resin and immediately collect 1 mL of eluted liquid into a 1.5 mL Eppendorf tube.
4. Determine the protein concentration of the desalted leaf total soluble protein extract by spectrophotometry using a commercial Bradford reagent [23], following the manufacturer's instructions. This requires a protein calibration curve using BSA (*see* **Note 14**).

3.3 PGLP1 Activity Measurement via a Coupled Reaction Using Recombinant At*GOX1*

1. Set up the spectrophotometer at 440 nm and make sure that the spectrophotometer cuvette chamber is at 30 °C.
2. Select the software function allowing the measurement of absorbance over time.
3. Pipette the appropriate volume (*see* **Note 15**) of reaction buffer (*see* **Note 9**) to a 1 mL plastic spectrophotometry cuvette. Place the cuvette in the spectrophotometer chamber and add in the following order: 40 μL of 20 mM *o*-dianisidine (*see* **Notes 1** and **4**), 20 μL of 100 U/mL HRP (*see* **Note 5**), and 40 μg purified recombinant *At*GOX1 (*see* **Note 10**). Mix with a disposable plastic spatula and wait 2 min to allow the reaction mix to equilibrate at 30 °C.
4. To test the proper functioning of the coupled-reaction enzymes (*see* **Note 16** and Fig. 2a), start by adding 10 μL of 0.5 M glycolic acid (*see* **Note 6** and Fig. 2a).
5. To measure a maximal PGLP activity, first add 40 μL of 25 mM 2-PG (1 mM final concentration; *see* **Notes** 7 and **17**) and wait 3 min (*see* **Note 18** and Fig. 2a) before starting the reaction by the addition of either 5 μg of purified PGLP (Fig. 2b) or 50 μg of desalted leaf total soluble protein (Fig. 2c, d) (*see* **Notes 11** and **15**).
6. Determine the initial velocity of the reaction by calculating the slope of the initial linear part of the absorbance curve (*see* Fig. 2b–d). To do this, use the Beer-Lambert equation with the extinction coefficient of 11.3/mM/cm at 440 nm for oxidized *o*-dianisidine and an optical path length of 1 cm. In our protocol using 5 μg of purified *At*PGLP1, the initial velocity (in μmol glycolate/min/mg PGLP protein) is calculated as follows ((A_{440nm}/min/5 μg/mL/PGLP1 protein)/11,300) × 1000 × (1000/5) (*see* **Note 20**).

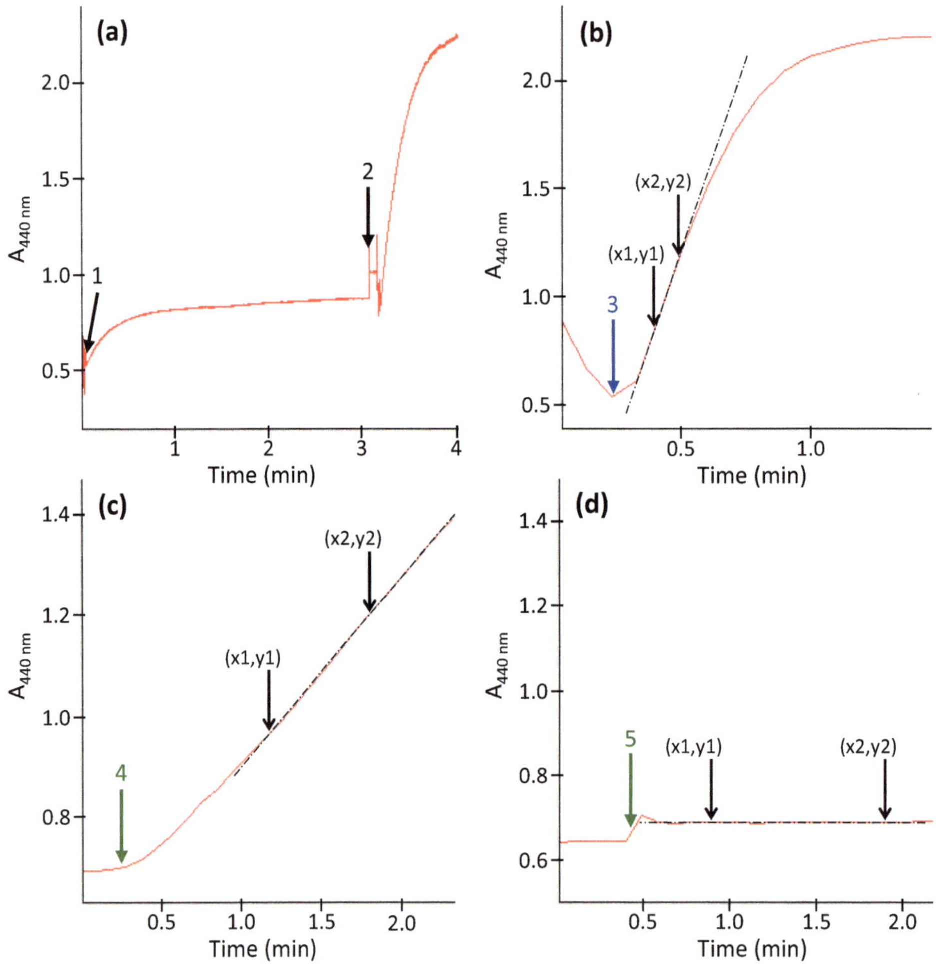

Fig. 2 Real-time spectrophotometer traces measuring *o*-dianisidine oxidation at 440 nm. Real-time spectrophotometer traces are shown in red. (**a**) Testing GOX-coupled reaction activity before starting PGLP activity measurements. The reaction mixture includes reaction buffer, 40 μL of 20 mM *o*-dianisidine (800 μM final), 20 μL of 100 U/mL HRP (2 U/mL final), and 40 μg purified recombinant *At*GOX1. At $t = 0$, the recording is started by the addition of 20 μL of 250 mM 2-PG (5 mM final) (black arrow n°1) and the reaction mix is quickly homogenized with a spatula (perturbation of the trace). The initial trace illustrates that the 2-PG solution contains a low amount of glycolate (that is 2-PG batch dependent). After 3 min, a plateau is reached and 10 μL of 0.5 M glycolate (5 mM final) is added (black arrow n°2) to determine that the GOX reaction is working efficiently. (**b–d**) Representative PGLP activity traces. Before starting recording, homogenized reaction mixtures containing reaction buffer, 40 μL of 20 mM *o*-dianisidine (800 μM final), 20 μL of 100 U/mL HRP (2u/mL final), 40 μg purified recombinant *At*GOX1, and 40 μL of 25 mM 2-PG (1 mM final) are incubated 3 min (to remove glycolate in the 2-PG solution). Recording of PGLP activity is started by the addition and mixing of (**b**) 5 μg of purified recombinant *At*PGLP1 (blue arrow n°3) or (**c**) 50 μg of soluble protein extract either from wild-type Arabidopsis rosette leaves (green arrow n°4) or (**d**) from a *pglp1* mutant [24] plant (green arrow n°5). In (**b–d**), ($x1$,$y1$) and ($x2$,$y2$) indicate time points on the trace that can be used to calculate the initial enzymatic velocity $((y2–y1)/(x2–x1))$. Panel (**d**) illustrates the specificity of the coupled reaction since no activity is detected in a total soluble protein extract from rosette leaves of the *pglp1* mutant

3.4 Determination of PGLP Kinetic Parameters

1. Prepare an adequate 2-PG concentration range using the 250 mM stock solution and 50 mM Tris–HCl, pH 8 (*see* **Note 17**).
2. Measure the initial PGLP activity for each 2-PG concentration by adding the appropriate volume of 2-PG solution (*see* **Note 19**) as previously described for purified PGLP (*see* Subheading 3.3).
3. Use the Michaelis–Menten equation to determine the kinetic parameters (*see* **Note 21**).

4 Notes

1. *o*-Dianisidine is classified as a carcinogenic, mutagenic, or reprotoxic (CMR) substance and it must be handled with caution. A mask and gloves should be worn when weighing the powder, and it is strongly recommended to wear gloves throughout all procedures. Disposable items containing *o*-dianisidine must be treated in accordance to laboratory rules for CMR products.
2. One tablet per 50 mL of buffer (extraction or reaction; *see* **Note 9**) is added and dissolved using a magnetic stirring bar and a magnetic stirrer.
3. Tris–HCl and $MgCl_2$ stock solutions for desalting and reaction buffers can be stored for several months at 4 °C. To prepare the 0.5 M Tris–HCl stock solution at pH 8, weigh 30.29 g of Tris–HCl, dissolve in 450 mL of ultrapure water, and adjust the pH to 8 using HCl. While concentrated HCl (12.1 N) can be used, it is best to adjust the pH with a lower concentrated HCl solution (1 N). Then adjust the volume to 500 mL in a 500 mL graduated cylinder with ultrapure water, homogenize and store in 250 mL glass bottles. To prepare the 0.5 M $MgCl_2$ stock solution, weigh 20.33 g of $MgCl_2.6H_2O$ and dissolve in 180 mL of ultrapure water before adjusting the volume with ultrapure water to 200 mL in a 250 mL graduated cylinder, homogenize and store in a 250 mL glass bottle.
4. *o*-Dianisidine is freshly prepared with ultrapure water and protected from light using aluminum foil. Since *o*-dianisidine is toxic, only a small amount is weighed and dissolved in the appropriate volume of ultrapure water to give a 20 mM solution. It is either kept on ice (during use) or conserved for a maximum of 1 week at 4 °C.
5. To prepare the HRP stock solution, weigh the appropriate amount of powder (batch dependent) and dissolve in ultrapure water to obtain a 1000 U/mL solution. Keep it protected from

light with aluminum foil. It can be stored at 4 °C for 1 month maximum. The 100 U/mL HRP working solution is freshly prepared from the 1000 U/mL HRP stock solution by dilution with ultrapure water. Keep it protected from light and on ice (during use).

6. Glycolic acid powder is dissolved in 50 mM Tris–HCl, pH 8. To prepare the 0.5 M glycolate solution, weigh 38 mg of glycolic acid and dissolve in 1 mL of 50 mM Tris–HCl, pH 8. The solution can be kept aliquoted at −20 °C for 1 year.
7. To obtain a 250 mM 2-PG stock solution, dissolve the entire 50 mg vial of 2-PG by adding 1281.8 μL of 50 mM Tris–HCl, pH 8 directly into the vial. Aliquot in small volumes (max. 200 μL) into 1.5 mL Eppendorf tubes to avoid freeze/thaw cycles. It can be kept at −20 °C for 1 year. When required, appropriate 2-PG diluted working solutions (*see* **Note 17**) are prepared with 50 mM Tris–HCl, pH 8 from the 250 mM 2-PG stock solution; they can be kept on ice (during use) and only frozen once at −20 °C.
8. Sample between 3 and 5 rosette leaves from 4- to 5-week-old short-day (8 h light/16 h dark) grown *Arabidopsis thaliana* plants at the middle of the photoperiod. Use forceps to remove leaves and quickly place them into a 2 mL Eppendorf tube containing 2 stainless steel beads. Sampling is carried out in the growth chamber to avoid changes in temperature and light conditions.
9. Extraction buffer is prepared by mixing 180 mL of ultrapure water with 20 mL of 0.5 M Tris–HCl, pH 8 and 4 tablets of antiprotease inhibitor cocktail (*see* **Note 2**). Reaction buffer is made by combining 44 mL of ultrapure water with 5 mL of 0.5 M Tris–HCl, pH 8 and 1 mL of 0.5 M $MgCl_2$ and one tablet of antiprotease inhibitor cocktail (*see* **Note 2**).
10. N-Terminal His-tagged *At*GOX1 recombinant protein is freshly purified by Ni-affinity chromatography and desalted (*see* Chap. 8) on the day of PGLP activity measurements due to its loss of activity when stored. All GOX containing solutions must be kept in the dark using aluminum foil due to the presence of flavin mononucleotide (FMN) as co-factor.
11. This assay is suitable for measuring either PGLP activity of leaf total soluble protein extracts (50 μg) or purified PGLP (5 μg; for example recombinant His-tagged *At*PGLP1; *see* Chap. 8).
12. From the freezer (−80 °C) to the ball mill machine, leaf samples are transported in liquid nitrogen.
13. After addition of extraction buffer, samples are kept on ice as often as possible.

14. A new calibration curve using BSA is required for each new bottle of commercial Bradford reagent. For an accurate measurement, protein concentrations must be in the correct protein detection window for the chosen Bradford reagent.
15. Before starting activity measurements, the required volumes of enzymes in the reaction mix (50 μg of leaf total soluble proteins or 5 μg of recombinant PGLP and 40 μg of *At*GOX1) are calculated and the reaction buffer volume is adjusted to a 1 mL final volume by also taking into account the 40 μL of 20 mM *o*-dianisidine (*see* **Notes 1** and **4**), 20 μL of 100 U/mL HRP (*see* **Note 5**) and the appropriate volume of 2-PG solution.
16. Each component of the coupled reaction assay must be in excess in order to measure PGLP activity correctly.
17. Appropriate 2-PG diluted working solutions are prepared with 50 mM Tris–HCl, pH 8. 25 mM 2-PG is prepared from the 250 mM 2-PG stock solution (*see* **Note 7**). 10 mM and 2.5 mM 2-PG are prepared from the fresh 25 mM 2-PG solution and 0.25 mM 2-PG is prepared from the fresh 2.5 mM 2-PG solution. They are kept on ice (during use) and only frozen once at −20 °C.
18. The 2-PG lithium salt may contain residual glycolate that should be metabolized before starting each PGLP activity measurement.
19. To determine PGLP kinetic parameters: Measure PGLP activity under an extended 2-PG concentration range (for example from 0.005 to 5 mM). Use 2-PG working or stock solutions (*see* **Notes 7** and **17**) so as to pipette between 10 and 40 μL into the spectrophotometer cuvette. Depending on the added volumes of 2-PG, the reaction buffer volume must be adjusted to maintain the same total liquid content (*see* also **Note 15**).
20. The calculated initial velocity obtained from the traces shown in Fig. 2b is expressed as A_{440nm}/min/5 μg/mL PGLP protein. To convert this into μmol of glycolate/min/mg PGLP protein, the Beer–Lambert equation is used. In order to correctly convert the units, several factors have to be applied. The Beer–Lambert equation units are as follows:

$$A_{440nm}\,[\text{no unit}] = {}_{440nm}[\text{L/mol/cm}]\,\text{l}[\text{cm}]\,\text{C}[\text{mol/L}]$$

(a) The extinction coefficient of 11.3/mM/cm at 440 nm for oxidized *o*-dianisidine is multiplied 1000 fold to obtain 11,300/μM/cm (or L/**μmol**/cm).

(b) To convert mL and L, the value is multiplied by a factor 1000.

(c) To convert 5 μg into 1 mg of PGLP protein, the value is divided by 5 and multiplied by 1000 to a value per **mg** of PGLP protein.

21. Determining these parameters can be done with specific software such as the "Kinetic Enzyme Wizard" of Sigma-Plot (Systat Software Inc.). This software provides several equations to characterize different types of enzymatic reactions.

Acknowledgments

We would like to thank Younès Dellero and Mathieu Jossier who developed the GOX activity measurement in the Metaboactions team and Herman Bauwe and Stefan Timm (University of Rostock) for kindly providing Arabidopsis *pglp1* mutant seeds. We would like to thank the French Agence Nationale de la Recherche (ANR) for financial support: ANR-10-LABX-0040-SPS, ANR-11-IDEX-0003-02, ANR-14-CE19-0015 REGUL3P. YD and PD received PhD fellowships from the French Education Ministry.

References

1. Allen KN, Dunaway-Mariano D (2004) Phosphoryl group transfer: evolution of a catalytic scaffold. Trends Biochem Sci 29:495–503
2. Burroughs AM, Allen KN, Dunaway-Mariano-D, Aravind L (2006) Evolutionary genomics of the HAD superfamily: understanding the structural adaptations and catalytic diversity in a superfamily of phosphoesterases and allied enzymes. J Mol Biol 361:1003–1034
3. Huang H, Pandya C, Liu C et al (2015) Panoramic view of a superfamily of phosphatases through substrate profiling. Proc Natl Acad Sci USA 112:1974–1983
4. Richardson KE, Tolbert NE (1961) Phosphoglycolic acid phosphatase. J Biol Chem 236: 1285–1290
5. Husic HD, Tolbert NE (1984) Anion and divalent cation activation of phosphoglycolate phosphatase from leaves. Arch Biochem Biophys 229:64–72
6. Norman EG, Colman B (1991) Purification and characterization of phosphoglycolate phosphatase from the cyanobacterium *Coccochloris peniocystis.* Plant Physiol 95:693–698
7. Bauwe H (2017) Measurement of enzyme activities. In: Fernie AR, Bauwe H, Weber APM (eds) Photorespiration: methods and protocols. Springer, New York, pp 31–50
8. Rose ZB (1981) Phosphoglycolate phosphatase from human red blood cells. Arch Biochem Biophys 208:602–609
9. Husic HD, Tolbert NE (1985) Properties of phosphoglycolate phosphatase from *Chlamydomonas reinhardtii* and *Anacystis nidulans.* Plant Physiol 79:394–399
10. Rose ZB, Grove DS, Seal SN (1986) Mechanism of activation by anions of phosphoglycolate phosphatases from spinach and human red blood cells. J Biol Chem 261:10996–11002
11. Kim Y, Yakunin AF, Kuznetsova E et al (2004) Structure- and function-based characterization of a new phosphoglycolate phosphatase from *Thermoplasma acidophilum.* J Biol Chem 279: 517–526
12. Segerer G, Hadamek K, Zundler et al (2016) An essential developmental function for murine phosphoglycolate phosphatase in safeguarding cell proliferation. Sci Rep 6:1–13
13. Calkins VP (1943) Microdetermination of glycolic and oxalic acids. Ind Eng Chem Res 15: 762–763
14. Ting IP, Osmond CB (1973) Photosynthetic phosphoenolpyruvate carboxylases: characteristics of alloenzymes from leaves of C_3 and C_4 plants. Plant Physiol 51:439–447

15. Echevarria C, Pacquit V, Bakrim N et al (1994) The effect of pH on the covalent and metabolic control of C4 phosphoenolpyruvate carboxylase from Sorghum leaf. Arch Biochem Biophys 315:425–430
16. Zhang Y, Foster JM, Kumar S et al (2004) Cofactor-independent phosphoglycerate mutase has an essential role in *Caenorhabditis elegans* and is conserved in parasitic nematodes. J Biol Chem 279:37185–37190
17. Duminil P, Davanture M, Oury C et al (2021) *Arabidopsis thaliana* 2,3-bisphosphoglycerate-independent phosphoglycerate mutase 2 activity requires serine 82 phosphorylation. Plant J 107:1478–1489
18. Dellero Y, Jossier M, Glab N et al (2016) Decreased glycolate oxidase activity leads to altered carbon allocation and leaf senescence after a transfer from high CO_2 to ambient air in *Arabidopsis thaliana*. J Exp Bot 67:3149–3163
19. Macheroux P, Massey V, Thiele DJ, Volokita M (1991) Expression of spinach glycolate oxidase in *Saccharomyces cerevisiae*: purification and characterization. Biochemistry 30:4612–4619
20. Stenberg K, Lindqvist Y (1996) High-level expression, purification, and crystallization of recombinant Spinach glycolate oxidase in *Escherichia coli*. Protein Expr Purif 8:295–298
21. Dellero Y, Mauve C, Boex-Fontvieille E et al (2015) Experimental evidence for a hydride transfer mechanism in plant glycolate oxidase catalysis. J Biol Chem 290:1689–1698
22. Launay A, Jolivet S, Clément G et al (2022) DspA/E-triggered non-host resistance against *E. amylovora* depends on the Arabidopsis GLYCOLATE OXIDASE 2 gene. Int J Mol Sci 23:4224. https://doi.org/10.3390/ijms23084224
23. Bradford MM (1976) A rapid and sensitive method for the quantitation of microgram quantities of protein utilizing the principle of protein-dye binding. Anal Biochem 72:248–254
24. Schwarte S, Bauwe H (2007) Identification of the photorespiratory 2-phosphoglycolate phosphatase, PGLP1, in *Arabidopsis*. Plant Physiol 144:1580–1586

Chapter 4

Spectrophotometric Assays for Measuring Photorespiratory Glutamate:Glyoxylate and Serine:Glyoxylate Aminotransferase Reactions

Mair C. Edwards and Aaron H. Liepman

Abstract

Glutamate:glyoxylate aminotransferase (GGAT; EC 2.6.1.4) and serine:glyoxylate aminotransferase activities (SGAT; EC 2.6.1.45) are central photorespiratory reactions within plant peroxisomes. Both enzymatic reactions convert glyoxylate, a product of glycolate oxidase, to glycine, a substrate of the mitochondrial glycine decarboxylase complex. The GGAT reaction uses glutamate as an amino group donor and also produces α-ketoglutarate, which is recycled to glutamate in plastids by ferredoxin-dependent glutamate synthase. Using serine, a product of mitochondrial serine hydroxymethyltransferase, as an amino group donor, the SGAT reaction also produces hydroxypyruvate, a substrate of hydroxypyruvate reductase. The activities of these photorespiratory aminotransferases can be measured using indirect, coupled, spectrophotometric assays, detailed herein.

Key words Photorespiration, Spectrophotometric assay, Glutamate:glyoxylate aminotransferase, Serine:glyoxylate aminotransferase, NADH, Enzyme kinetics

1 Introduction

Within the photorespiratory pathway, two types of aminotransferases catalyze transaminations using glyoxylate as an amino acceptor within plant peroxisomes [1, 2]. Glutamate glyoxylate aminotransferase (GGT) proteins catalyze the glutamate:glyoxylate aminotransferase (GGAT) reaction (EC 2.6.1.4; Reaction 1) [3, 4].

$$\text{Glutamate} + \text{Glyoxylate} \rightarrow \alpha - \text{ketoglutarate} + \text{Glycine} \quad \text{(Reaction 1)}$$

The serine:glyoxylate aminotransferase (SGAT) reaction (EC 2.6.1.45; Reaction 2) is catalyzed by alanine glyoxylate aminotransferase 1 (AGT1) homologs [5].

Berkley J. Walker (ed.), *Photorespiration: Methods and Protocols*, Methods in Molecular Biology, vol. 2792,
https://doi.org/10.1007/978-1-0716-3802-6_4,

$$\text{Serine} + \text{Glyoxylate} \rightarrow \text{Hydroxypyruvate} + \text{Glycine} \qquad \text{(Reaction 2)}$$

Both of these enzymes catalyze reactions using additional donor and acceptor substrates [2, 6–8].

Because products of the GGAT and SGAT reactions are substrates for NADH-dependent coupling enzymes, both reactions can be measured in vitro using indirect spectrophotometric assays. GGAT activity is measured by coupling the reductive amination by the coupling enzyme glutamic dehydrogenase (EC 1.4.1.2; Reaction 3) of α-ketoglutarate, a product of the GGAT reaction.

$$\alpha - \text{ketoglutarate} + NH_3 + \text{NADH} + H^+ \rightarrow \text{Glutamate} + \text{NAD}^+ + H_2O \qquad \text{(Reaction 3)}$$

SGAT activity is measured by coupling reduction by the coupling enzyme hydroxypyruvate reductase (EC 1.1.1.81; Reaction 4) of hydroxypyruvate, a product of the SGAT reaction.

$$\text{Hydroxypyruvate} + \text{NADH} + H^+ \rightarrow \text{D} - \text{glycerate} + \text{NAD}^+ \qquad \text{(Reaction 4)}$$

Since the coupling enzymes used for GGAT and SGAT assays both use NADH as cofactor, these glyoxylate aminotransferase activities are measured by following the consumption of NADH at 340 nm. These assays can be conducted using cuvettes [1, 9, 10] or scaled down for 96-well microplates, facilitating simultaneous measurements of many samples using a plate-reader UV spectrophotometer. These assays can be used for measuring activity of glyoxylate aminotransferase proteins purified from plant extracts (e.g., [2, 3, 6]) or purified recombinant proteins (e.g., [3, 5, 8]). Here, we describe conditions for microplate scaled assays for measuring activity of glyoxylate aminotransferases.

2 Materials

Unless otherwise noted, all solutions should be prepared using analytical grade reagents dissolved in ultrapure water (18 MΩ * cm at 25 °C). When disposing of waste, adhere to appropriate waste disposal guidelines. Except where noted, stock solutions can be prepared ahead of time, stored frozen (−20 °C), and thawed when needed.

2.1 Materials and Stocks Used for Glutamate:Glyoxylate Aminotransferase Assays

1. Temperature-controlled UV spectrophotometer (standard or plate-reader), set to read absorbance at 340 nm.
2. UV-transparent cuvettes (disposable or quartz) or flat-bottomed 96-well plates (*see* **Note 1**).
3. 1 M potassium phosphate buffer, pH 7.5 (*see* **Note 2**).
4. 1.7 mM NADH, dissolved in 10 mM NaOH (*see* **Note 3**).
5. 50 mM pyridoxal-5′-phosphate (PLP) (*see* **Note 4**).
6. 100 mM ammonium chloride.
7. 100 mM glutamic acid (*see* **Note 5**).
8. 10 mM glyoxylate.
9. Glutamic dehydrogenase (*see* **Note 6**).
10. GGAT enzyme dilution buffer: 50 mM Tris–HCl pH 8, 10% glycerol, 0.2 mM PLP.

2.2 Materials and Stocks Used for Serine:Glyoxylate Aminotransferase Assays

1. Temperature-controlled UV spectrophotometer (standard or plate-reader), set to read absorbance at 340 nm.
2. UV-transparent cuvettes (disposable or quartz) or 96-well plates.
3. 1 M HEPES-KOH buffer, pH 7 (*see* **Note 7**).
4. 1.7 mM NADH, dissolved in 10 mM NaOH (*see* **Note 3**).
5. 50 mM pyridoxal-5′-phosphate (PLP) (*see* **Note 4**).
6. 100 mM serine.
7. 10 mM glyoxylate.
8. Hydroxypyruvate reductase (*see* **Notes 8** and **9**).
9. SGAT enzyme dilution buffer: 50 mM Tris–HCl pH 8, 100 mM NaCl, 10% glycerol, 0.1 mM PLP.

3 Methods

These activity assays can be conducted using cuvettes (1 mL per sample) or 96-well plates (200 μL per sample). Prepare an assay cocktail of sufficient volume for the desired quantity of samples and replicates. We typically conduct assays in triplicate. When measuring K_m of a substrate, exclude this substrate from the assay cocktail and add it directly to samples, at a range of concentrations, to initiate reactions (*see* **Note 10**). Absorbance data (OD_{340}) are collected in kinetics mode. It is best to use a quantity of enzyme that yields a linear OD_{340} decrease over a 4–6 min period (*see* **Note 11**) and remains linear while approximately doubling in activity when twice the quantity of enzyme is used.

Table 1
GGAT assay components

Reagent and final concentration	Stock concentration	Stock volume per sample	Volume for 10 reactions (96-well format)
Water	N/A	103 μL	1030 μL
50 mM potassium phosphate pH 7.5	1 M	10 μL	100 μL
3.1 mM ammonium chloride	100 mM	6.2 μL	62 μL
0.1 mM PLP	50 mM	0.4 μL	4 μL
0.17 mM NADH	1.7 mM	20 μL	200 μL
2.2 U/mL glutamic dehydrogenase	1 U/μL	0.44 μL	4.4 μL
10 mM glutamate	100 mM	20 μL	200 μL
5 mM glyoxylate	100 mM	20 μL	200 μL

Reactions are initiated by adding 20 μL of diluted enzyme (or water, control) to bring total sample volume to 200 μL

3.1 Glutamate: Glyoxylate Aminotransferase Assay

1. Prepare assay cocktail as indicated in Table 1, mix thoroughly, and store at room temperature.
2. Dilute enzyme in GGAT enzyme dilution buffer. We find that ~75 ng of purified recombinant Arabidopsis GGT1 protein per assay sample yields linear readings within the assay duration (*see* **Note 12**). Aliquot diluted enzyme into 8-tube PCR tube strips to facilitate multichannel pipetting.
3. Set spectrophotometer to measure OD_{340} in kinetics mode at 25 °C (*see* **Note 13**), collecting readings at ~15 s intervals.
4. Transfer 180 μL of assay cocktail into designated wells of a 96-well plate.
5. Using an 8-channel pipette, transfer 20 μL of GGAT enzyme dilution (or GGAT dilution buffer without enzyme for no enzyme control reactions) from PCR strip tubes into wells of the 96-well plate and mix by pipetting up and down several times.
6. Immediately initiate kinetics readings. Save/export data to a spreadsheet for calculations.
7. When calculating enzyme activity, use only data within the linear portion of the assay. Subtract the average absorbance change of blank samples (lacking enzyme or one substrate, as appropriate) from the average absorbance change of whole samples. Calculate the quantity of NADH consumed per minute using the molar extinction coefficient for NADH at 340 nm (6220 M^{-1} cm^{-1}) with correction for path length, as necessary

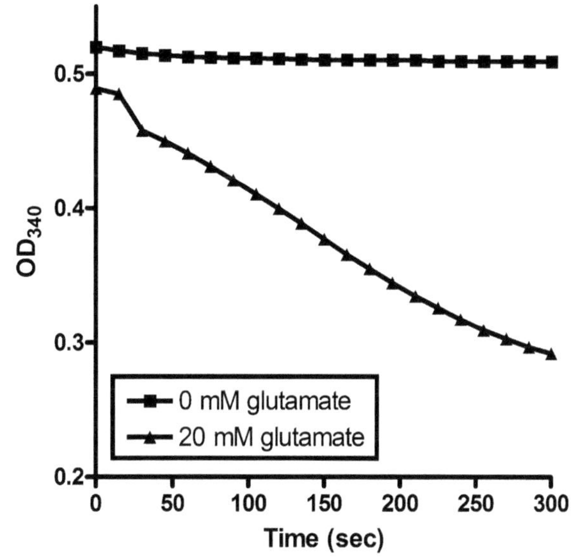

Fig. 1 Representative kinetics data from a coupled spectrophotometric assay of GGAT activity of a GGT1 protein from *Cyanidioschyzon merolae*

(*see* **Note 14**). Specific activity (U/mg) may be calculated if using a purified protein of known concentration. A plot showing typical kinetics data for a GGAT assay is shown in Fig. 1.

3.2 Serine:Glyoxylate Aminotransferase Assay

1. Prepare assay cocktail as indicated in Table 2, mix thoroughly, and store at room temperature.
2. Dilute enzyme preparation and HPR1 coupler enzyme separately into SGAT enzyme dilution buffer. We find that ~125 ng (10 μL diluted to 12.5 ng/μL) of purified recombinant Arabidopsis AGT1 protein per assay sample yields linear readings over the assay duration (*see* **Note 12**). An excess of Arabidopsis HPR1 is included as coupler enzyme (~ 0.01875 U/μL in the assay—this is equivalent to ~45 ng of purified recombinant AtHPR1). Aliquot diluted enzymes into 8-tube PCR tube strips to facilitate multichannel pipetting.
3. Set spectrophotometer to measure OD_{340} in kinetics mode at 25 °C (*see* **Note 13**), collecting readings at ~15 s intervals.
4. Transfer 180 μL of assay cocktail per well of a 96-well plate.
5. Using an 8-channel pipette, transfer 10 μL of AGT1 enzyme dilution (or SGAT dilution buffer without enzyme for no enzyme control reactions; *see* **Note 15**) from PCR strip tubes into wells of the 96-well plate and mix by pipetting up and down several times.

Table 2
SGAT assay components

Reagent and final concentration	Stock concentration	Stock volume per sample	Volume for 10 reactions (96-well format)
Water	N/A	123.6 μL	1236 μL
70 mM HEPES-KOH pH 7	1 M	14 μL	140 μL
0.1 mM PLP	50 mM	0.4 μL	4 μL
0.17 mM NADH	1.7 mM	20 μL	200 μL
10 mM serine	100 mM	20 μL	200 μL
1 mM glyoxylate	100 mM	2 μL	20 μL

Reactions are initiated by adding 10 μL of diluted HPR coupler enzyme and 10 μL of enzyme sample (or water, control) to bring total sample volume to 200 μL

6. Using an 8-channel pipette, transfer 10 μL of HPR1 coupler enzyme dilution from PCR strip tubes into wells of the 96-well plate and mix by pipetting up and down several times.
7. Immediately initiate kinetics readings. Save/export data to a spreadsheet for calculations.
8. When calculating enzyme activity, use only data within the linear range of the assay. Subtract the average absorbance change of blank samples (controls containing HPR1 coupler enzyme and lacking AGT1) from the average absorbance change of samples containing AGT1 and HPR1 coupler enzyme. Calculate the quantity of NADH consumed per minute using the molar extinction coefficient for NADH at 340 nm (6220 L/M cm) with correction for path length, as necessary (*see* **Note 14**). Specific activity (U/mg) may be calculated if using purified protein of known concentration. A plot showing typical kinetics data for an SGAT assay is shown in Fig. 2.

4 Notes

1. We use Corning 9017 96-well plates, though others can be used.
2. 100 mM HEPES-KOH buffer (pH 7.5) also has been used [10].
3. We prepare a fresh NADH stock and use it the same day for assays.
4. PLP is a co-factor required by aminotransferases. The addition of NaOH improves the dissolution of PLP. When preparing a stock solution of PLP, we add 1 M NaOH, dropwise, and mix thoroughly until all powder has dissolved.

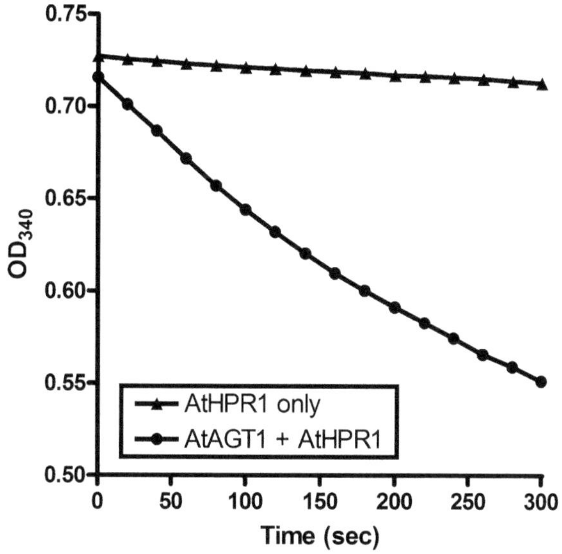

Fig. 2 Representative kinetics data from a coupled spectrophotometric assay of SGAT activity of the AGT1 protein from *Arabidopsis thaliana*. Because the coupling enzyme (AtHPR1) also exhibits glyoxylate reductase activity, SGAT reaction rates are calculated by subtracting the rate of control reactions (AtHPR1 only) from reactions containing AtAGT1 and AtHPR1

5. Dissolution of glutamic acid is facilitated by heating (65 °C). After storage at −20 °C, some precipitation may occur. Briefly heating the solution to 65 °C and thorough mixing can be used to re-dissolve precipitated materials.
6. Ammonium sulfate suspension.
7. Phosphate buffers inhibit SGAT activity [1].
8. There are various commercial sources of recombinant hydroxypyruvate reductase, though we have not tested these as coupler enzyme for SGAT assays. We use purified recombinant Arabidopsis HPR1 protein [11].
9. When using purified recombinant AtHPR1 as a coupler enzyme for SGAT assays, we include ~3.75 U per 200 μL assay sample (~0.0188 U/μL). This is equivalent to ~45 ng of purified protein per assay.
10. To determine K_m values for amino acid substrates of photorespiratory aminotransferases, a range of 0.25–20 mM is appropriate. To determine K_m values for glyoxylate, a range of 25 μM to 2.5 mM is appropriate. For kinetics assays, prepare a modified cocktail lacking the amino acid or glyoxylate substrate of interest and add the appropriate dilution of the missing substrate directly to samples.
11. We aim for a total OD_{340} change of ~0.1 AU.

12. It may be necessary to sample various dilutions of an enzyme preparation to determine an appropriate quantity to achieve assay linearity over a 4–6 min period. It is important to verify that enzyme activity varies proportionally to the quantity of enzyme used for assays. If doubling the amount of enzyme roughly doubles the amount of activity observed, this indicates that an excess of coupler enzyme is present.
13. When conducting assays at other temperatures (e.g., for kinetics measurements at different temperatures), preincubate the 96-well plate and assay cocktail at the desired temperature before conducting assays.
14. To calculate enzyme activity for 200 μL assays conducted in 96-well plates, the path length is ~0.571 cm.
15. Because HPR proteins also possess glyoxylate reductase activity, it is necessary to subtract the rate of NADH consumption in reactions containing HPR coupler enzyme but lacking AGT1 from the rate of samples containing both AGT1 and HPR coupler enzyme.

Acknowledgments

This work was supported, in part, by a grant from the National Science Foundation (NSF/IOS award #2030295) to A.H.L.

References

1. Rehfeld DW, Tolbert NE (1972) Aminotransferases in peroxisomes from spinach leaves. J Biol Chem 247:4803–4811
2. Nakamura Y, Tolbert NE (1983) Serine: glyoxylate, alanine:glyoxylate, and glutamate:glyoxylate aminotransferase reactions in peroxisomes from spinach leaves. J Biol Chem 258:7631–7638
3. Liepman AH, Olsen LJ (2003) Alanine aminotransferase homologs catalyze the glutamate: glyoxylate aminotransferase reaction in peroxisomes of *Arabidopsis*. Plant Physiol 131:215–227. https://doi.org/10.1104/pp.011460
4. Igarashi D, Miwa T, Seki M, Kobayashi M, Kato T, Tabata S, Shinozaki K, Ohsumi C (2003) Identification of photorespiratory glutamate:glyoxylate aminotransferase (GGAT) gene in *Arabidopsis*. Plant J 33:975–987
5. Liepman AH, Olsen LJ (2001) Peroxisomal alanine:glyoxylate aminotransferase (AGT1) is a photorespiratory enzyme with multiple substrates in *Arabidopsis thaliana*: AGT1, a peroxisomal photorespiratory transaminase. Plant J 25:487–498. https://doi.org/10.1046/j.1365-313x.2001.00961.x
6. Kendziorek M, Paszkowski A (2008) Properties of serine:glyoxylate aminotransferase purified from *Arabidopsis thaliana* leaves. ABBS 40:102–110. https://doi.org/10.1111/j.1745-7270.2008.00383.x
7. Noguchi T, Fujiwara S (1982) Development of glutamate:glyoxylate aminotransferase in the cotyledons of cucumber (*Cucumis sativus*) seedlings. Biochem J 201:209–214. https://doi.org/10.1042/bj2010209
8. Zhang Q, Lee J, Pandurangan S, Clarke M, Pajak A, Marsolais F (2013) Characterization of *Arabidopsis* serine:glyoxylate aminotransferase, AGT1, as an asparagine aminotransferase. Phytochemistry 85:30–35. https://doi.org/10.1016/j.phytochem.2012.09.017
9. Huang AHC, Liu KDF, Youle RJ (1976) Organelle-specific isozymes of aspartate-α-ketoglutarate transaminase in spinach leaves. Plant Physiol 58:110–113

10. Bauwe H (2017) Measurement of enzyme activities. In: Fernie AR, Bauwe H, Weber APM (eds) Photorespiration. Springer, New York, pp 31–50

11. Timm S, Nunes-Nesi A, Pärnik T, Morgenthal K, Wienkoop S, Keerberg O, Weckwerth W, Kleczkowski LA, Fernie AR, Bauwe H (2008) A cytosolic pathway for the conversion of hydroxypyruvate to glycerate during photorespiration in *Arabidopsis*. Plant Cell 20:2848–2859. https://doi.org/10.1105/tpc.108.062265

Chapter 5

Mitochondrial Dihydrolipoamide Dehydrogenase (mtLPD1): Expression, Purification, Activity, and Redox Regulation

Stefan Timm, Kathrin Jahnke, Maike Cosse, and Jennifer Selinski

Abstract

Mitochondrial dihydrolipoamide dehydrogenase (mtLPD1) is a central enzyme in primary carbon metabolism, since its function is required to drive four multienzymes involved in photorespiration, the tricarboxylic acid (TCA) cycle, and the degradation of branched-chain amino acids. However, in illuminated, photosynthesizing tissue a vast amount of mtLPD1 is necessary for glycine decarboxylase (GDC), the key enzyme of photorespiration. In light of the shared role, the functional characterization of mtLPD1 is necessary to understand how the three pathways might interact under different environmental scenarios. This includes the determination of the biochemical properties and all potential regulatory mechanisms, respectively. With regards to the latter, regulation can occur through multiple levels including effector molecules, cofactor availability, or posttranslational modifications (PTM), which in turn decrease or increase the activity of each enzymatic reaction. Gaining a comprehensive overview on all these aspects would ultimately facilitate the interpretation of the metabolic interplay of the pathways within the whole subcellular network or even function as a proof of concept for genetic engineering approaches. Here, we describe the typical workflow how to clone, express, and purify plant mtLPD1 for biochemical characterization and how to analyze potential redox regulatory mechanisms in vitro and *in planta*.

Key words Mitochondrial dihydrolipoamide dehydrogenase (mtLPD1), Recombinant expression, Affinity purification, Enzyme assay, Redox state, Redox regulation

1 Introduction

Mitochondrial dihydrolipoamide dehydrogenase (mtLPD1; EC 1.8.1.4) is a ubiquitous enzyme involved in the primary metabolism of aerobic organisms. It is structurally highly conserved, representing a homodimeric flavoenzyme of the pyridine nucleotide-disulfide oxidoreductase enzyme family. As such, mtLPD1 is an integral component of four key mitochondrial multienzymes associated with three metabolic routes including photorespiration, the tricarboxylic acid (TCA) cycle, and the degradation of branched-chain amino acids (BCAAs). In more detail, mtLPD1 represents the E3 subunit of the photorespiratory multienzyme

Berkley J. Walker (ed.), *Photorespiration: Methods and Protocols*, Methods in Molecular Biology, vol. 2792,
https://doi.org/10.1007/978-1-0716-3802-6_5,

glycine decarboxylase (GDC); the TCA-cycle enzymes pyruvate dehydrogenase complex (PDHC) and 2-oxoglutarate dehydrogenase complex (OGDC) as well as the branched-chain 2-oxoacid dehydrogenase complex (BCDHC). The mtLPD1 protein is functionally required in each of these multienzymes to reoxidize a protein-bound (E2, in case of PDHC, BCDHC, and BCDHC or H-protein, in case of GDC) lipoyl group that becomes reduced in course of the respective reaction cycles in a NAD^+-dependent manner [1–3].

Despite the integrative enzymatic function of mtLPD1, changes in mtLPD1 activity might, however, differentially impact each multienzyme and the associated metabolic pathway under changing environmental conditions. This is mainly because there are considerable differences in the carbon fluxes of the three pathways, ranging from very high (photorespiration; in illuminated leaves), medium (TCA cycle) to rather low (degradation of BCAAs). Thus, depending on the environment, mtLPD1 contributes to the maintenance of the mitochondrial redox state and, perhaps, those of the entire cell. The dominant role of mtLPD1 in illuminated leaves is associated with photorespiratory GDC. This statement is in line with the high requirements for photorespiratory flux under current environmental conditions and results from Arabidopsis overexpression mutants, both which demonstrate increased mtLPD1 availability strongly shapes photosynthetic and photorespiratory metabolism [4]. Physiological responses were also observed on both of the other metabolic pathways (TCA cycle and degradation of BCAAs) that rely on mtLPD1-dependent multienzymes; however, these effects are less severely pronounced compared to the impact on photorespiration.

Given the multifaced role of mtLPD1, it seems reasonable to hypothesize that proper regulation of its enzymatic activity could be a key mechanism to adjust the different pathway fluxes in response to fluctuating environments including changes in light, temperature, the CO_2/O_2 ratio, or water availability [5, 6]. For this reason, and to rationally engineer plant primary metabolism, it is important to unravel all possible levels of mtLPD1 regulation, including PTM. Interestingly, recent research has provided new insights into its multi-layer redox regulation. On the one hand, it was demonstrated that thioredoxins (TRXs) are involved in the regulation of mtLPD1 activity, in vitro and *in planta* [7, 8]. The current experimental evidence in this regard suggests that TRX seems to control the activation state of mtLPD1, preventing the mitochondria from overreduction due to excessive production of NADH. On the other hand, and in line with the above statement, mtLPD1 is inactivated by NADH. This inhibition represents another possibility to adjust the activity of the whole multienzymes to which mtLPD1 functionally contributes [9, 10], independent from direct TRX regulation modes. To rationally explain that several different layers of enzymatic regulation, including redox regulation exist, it is worth mentioning photorespiration, the TCA cycle, and the degradation of BCAAs operate in parallel in the same subcellular compartment. Therefore, it is necessary to mechanistically understand all regulatory modes of mtLPD1 and the conditions under which they occur.

Within this chapter, we provide a workflow for how mtLPD1 can be comprehensively characterized from plant species. The protocol includes mtLPD1 cloning, recombinant expression, affinity purification, in vitro enzyme activity measurements, and two methods to characterize potential redox regulatory mechanisms in vitro and *in planta*. Regulation of mtLPD1 is likely to occur also through other post-translational modifications (PTMs) that are, however, not in focus of this chapter and require further specific protocols.

2 Materials

All solutions are prepared using ultrapure water (obtained via purification of deionized water, to attain a sensitivity of 18 MΩ-cm at 25 °C) and chemicals of analytical grade. Unless stated otherwise, all aqueous solutions are prepared at room temperature and stored in a fridge (buffers) or at −20 °C (enzymatic substrates) until further use. The equipment mentioned are recommendations but could be replaced by comparable devices with the same specifications. Parts of the experimental workflow and results thereof are schematically illustrated in Figs. 1, 2 and 3, while all required materials and methods to perform each step are specified in the text.

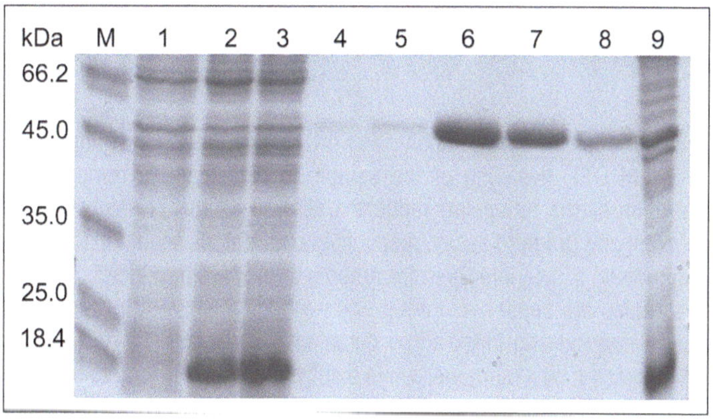

Fig. 1 SDS gel image of the mtLPD1 purification procedure. Selected fractions of the mtLPD1 purification process were subjected to SDS-page analysis to visualize the purified protein. M—protein standard, 1—uninduced cells, 2—total lysate, 3—flow through, 4—wash (20 mM imidazole), 5—wash (80 mM imidazole), 6—elution 1, 7—elution 2, 8—elution 3, 9—insoluble fraction

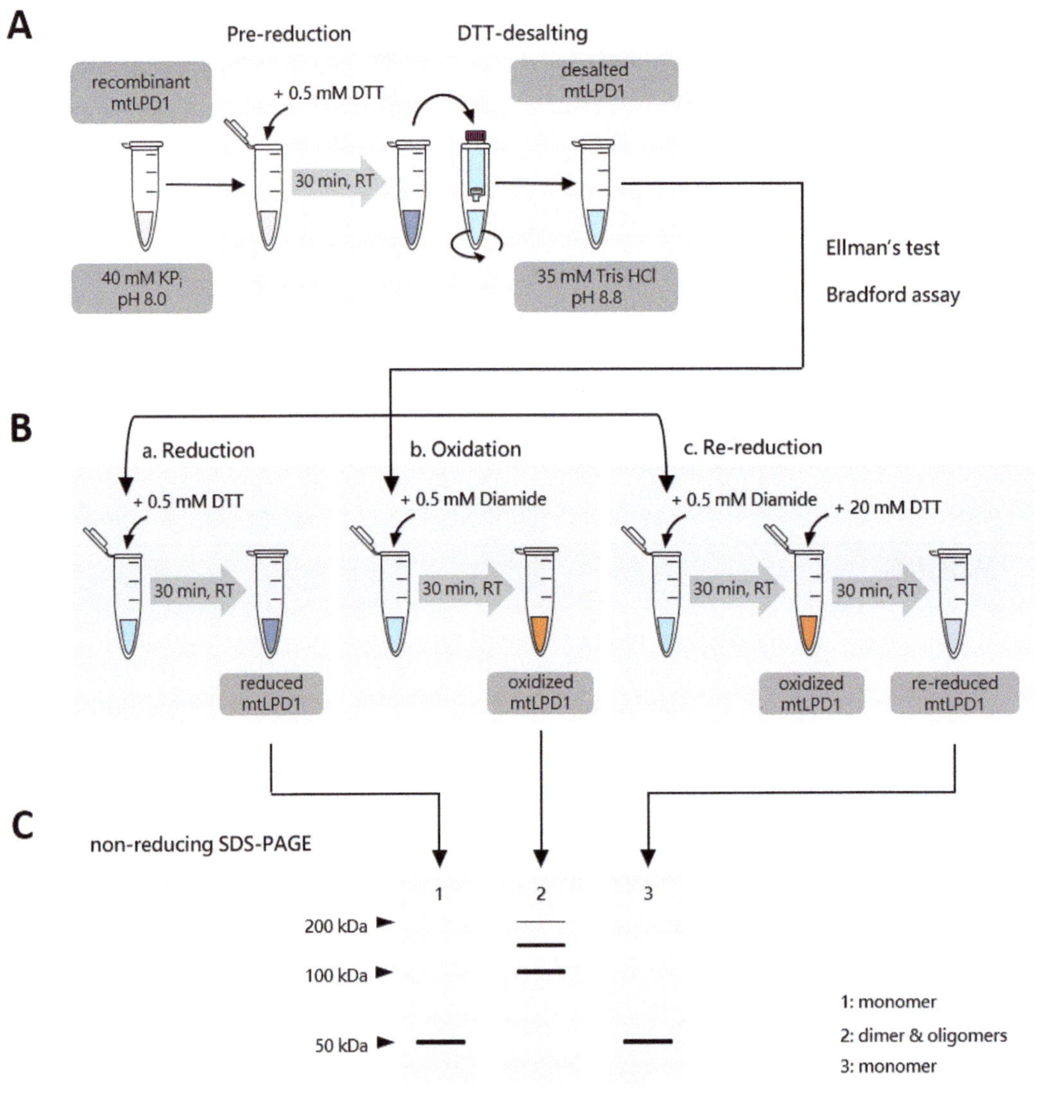

Fig. 2 Scheme for the in vitro analysis of mtLPD1 redox modifications. (**a**) Following prereduction of mtLPD1 with 0.5 mM DTT, desalting of the sample is necessary to remove excess reductant. Subsequently, the concentration of the remaining reductant as well as the protein concentration need to be determined via Ellman's test and Bradford assay, respectively (for details see text). (**b**) Next, prereduced and desalted mtLPD1 (~15 μg protein) is split into three treatments to prepare a reduced, an oxidized, and a re-reduced sample. (**c**) These samples are separated through non-reducing SDS-PAGE. Subsequently, the protein is visualized via Coomassie staining and followed by destaining of the gel (for details see text). Under reducing conditions, mtLPD1 is present as a monomer while oxidizing conditions lead to the formation of dimers or even oligomers of higher order. However, re-reduction of mtLPD1 leads back to the monomeric form showing the reversibility of mtLPD1's redox modification

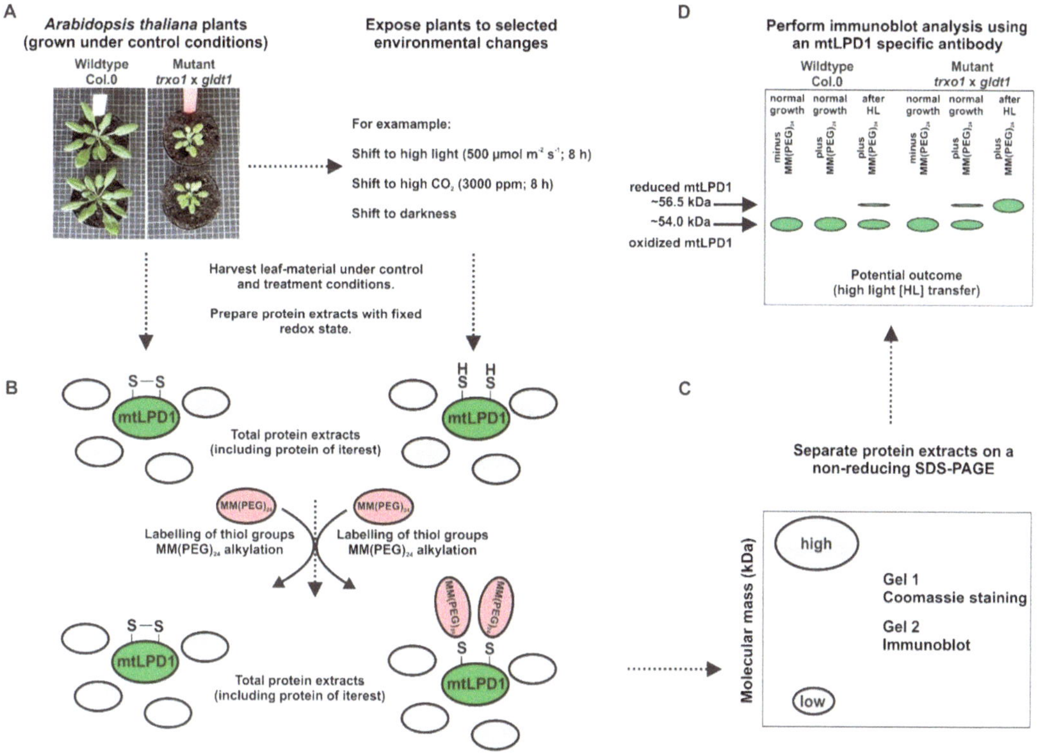

Fig. 3 Scheme for the *in planta* determination of the mtLPD1 redox state. (**a**) Arabidopsis wild-type and mutant plants (e.g., *trxo1xgldt1* double mutant; [7]) are grown under standard growth conditions to stage 5.1 [14]. Subsequently, a subset of plants is exposed to an environmental change of interest, potentially affecting the protein redox state, while another subset of plant remains under control conditions. Leaf material is harvested under both conditions at the same timepoint of the diurnal cycle wherever possible and immediately frozen in liquid nitrogen. (**b**) Protein extracts are prepared under conditions fixing the *in planta* redox state followed by labeling of the thiol groups through *in vitro* alkylation using the protein shifter MM(PEG)$_{24}$ (see text for details). (**c**) Protein extracts (use non-alkylated control samples and alkylated samples of plants grown under standard conditions or following environmental treatments) are separated through non-reducing SDS-PAGE twice. The first gel is used for Coomassie staining (loading control), while the other is subjected to immunoblotting analysis (**d**) using a mtLPD1-specific antibody. Compare mtLPD1 migration patterns of the different genotypes and environmental conditions and determine the ratio of oxidized and reduced mtLPD1 wherever appropriate. Each MM(PEG)$_{24}$ molecule shifts the total mtLPD1 protein size to about 1.24 kDa

2.1 PCR-Amplification and Cloning of mtLPD1

2.1.1 PCR-Amplification of mtLPD1

1. PCR cycler, pipettes, and consumables.
2. Proof-reading DNA-polymerase and PCR chemicals specific for the enzyme.
3. Oligonucleotide stock solutions (genomic scale, e.g., 100 µM—dilute 1:20 prior PCR) specific for the mtLPD1 to be analyzed, with specific restriction enzyme recognition sites according to the cloning strategy.
4. Template cDNA, synthesized from plant total RNA following standard procedures [11].
5. DNA size standard (e.g., 1 kB ladder).

6. Agarose gel electrophoresis equipment and power supply.
7. Running buffer: 1× Tris-acetic acid-EDTA (TAE), prepare a 50× TAE stock solution, 2 M Tris–HCl (pH. 8.0), 50 mM EDTA, 5.5% (v/v) acetic acid.
8. Agarose gel: The standard concentration for agarose gels used is 1% in 1× TAE buffer but might be adapted to the expected fragment sizes to be separated.
9. Ethidium bromide stock solution: 2.45 μM ethidium bromide in sterile water, use in a final concentration of 1:40,000 (*see* **Note 1**).
10. DNA loading dye: 10 mM Tris–HCl (pH 7.5), 1 mM EDTA, 0.2% (v/v) bromophenol blue, 30% (v/v) glycerol.
11. UV-based visualization device for PCR products.
12. Gel extraction or PCR product purification kit.
13. Heating blocks for microtubes.
14. Microcentrifuge.

2.1.2 Cloning and Sequence Verification of mtLPD1

1. Cloning vector for blunt-end PCR products.
2. *Escherichia coli* competent cells (strain DH5α or TOP10) for plasmid amplification.
3. Heating blocks for microtubes and ice bath.
4. Sterile test tubes, flasks, and incubator with temperature control with shaker to facilitate bacterial growth.
5. Luria-Bertani (LB) broth media and LB-agar, antibiotics specific for the vectors used (pJET1.2; ampicillin [amp], final concentration 200 μg mL^{-1}).
6. Plasmid DNA isolation kit.
7. Microcentrifuge.
8. Restriction enzymes (according to the cloning strategy).
9. Protein expression vector (pET28a) and competent cells (strain BL21gold) suitable for optimized protein expression.

2.2 Heterologous Expression and Affinity Purification of Recombinant mtLPD1

2.2.1 Expression of Recombinant mtLPD1

1. Incubator and shaker set to 37 °C (*see* **Note 2**).
2. Petri dishes, test tubes, Erlenmeyer flasks, LB-agar, and LB-medium.
3. Antibiotics and inductor specific for the expression vector used (pET28a: kanamycin [kan]: final concentration 100 μg mL^{-1}), isopropyl β-D-thiogalactoside (IPTG: final concentration 1 mM).
4. Spectrophotometer and semi-microplastic cuvettes.
5. Microcentrifuge.
6. Refrigerated ultracentrifuge (capable of reaching 20,000 *g*) and centrifuge tubes (suitable for volumes up to 500 mL).

2.2.2 Affinity Purification of Recombinant mtLPD1

1. Sonicator.
2. Lysozyme (final concentration 1 mg mL^{-1}; ~23,000 U/mg).
3. Columns (small-scale preparation), suitable for immobilized metal ion affinity chromatography (IMAC) of his-tagged proteins.
4. Equilibration buffer: 20 mM Tris–HCl (pH 8.0) containing 300 mM NaCl.
5. Homogenization buffer: 20 mM Tris–HCl (pH 8.0) containing 300 mM NaCl, and 20 mM imidazole.
6. Washing buffer: 20 mM Tris–HCl (pH 8.0) containing 300 mM NaCl and 80 mM imidazole (*see* **Note 3**).
7. Elution buffer: 20 mM Tris–HCl (pH 8.0) containing 300 mM NaCl and 300 mM imidazole.
8. 20% (v/v) ethanol.
9. Desalting columns (Sephadex with G-25 Medium) for NaCl and imidazole removal.

2.2.3 Protein Quantification

1. Spectrophotometer with selectable wavelength.
2. Pipettes and semimicro plastic cuvettes.
3. Bradford reagent for protein quantification [12] (*see* **Note 4**).

2.2.4 Sodium Dodecyl Sulfate-Polyacrylamide Gel Electrophoresis (SDS-PAGE)

1. SDS-PAGE equipment and power supply.
2. Shaker and heating blocks for microtubes.
3. 0.5 M Tris–HCl (pH 6.8) and 1.5 M Tris–HCl (pH 8.8) buffer.
4. Acrylamide stock solution (e.g., Rotiphorese Gel®30 ready to use [37.5:1]; Carl Roth).
5. 10% (v/v) SDS, in distilled water.
6. 10% (v/v) ammonium persulfate, in distilled water.
7. TEMED (*N*,*N*,*N*,*N*′-tetramethyl-ethylenediamine).
8. SDS electrophoresis buffer: 1× electrophoresis buffer, prepare a 10× SDS electrophoresis buffer stock solution, 280 mM Tris–HCl, 1.92 M glycine, 1% (v/v) SDS.
9. SDS-PAGE size standard (Unstained Protein Molecular Weight Marker; for protein sizes up to 100 kDa).
10. 3× Laemmli-buffer: 0.5 M Tris–HCl (pH 6.8), 30% (v/v) glycerol, 15% (v/v) β-mercaptoethanol, 0.15% (v/v) bromophenole blue, and 6% (v/v) SDS, in distilled water.
11. Coomassie staining solution: 0.3 mM Coomassie brilliant blue, in 200 mL distilled water, 250 mL methanol, and 50 mL acetic acid.

12. Destaining solution: 500 mL distilled water; 400 mL methanol; 100 mL acetic acid.
13. Fixation solution: 7% (v/v) acetic acid.

2.3 Enzyme Activity Measurements of Recombinant mtLPD1 (See Note 5)

1. Spectrophotometer, temperature-controlled with selectable wavelength (set to 25 °C, 340 nm), preferentially with a multiple cuvette holder (*see* **Note 6**).
2. Standard semi-micro UV-transparent (>320 nm) disposable plastic cuvettes.
3. Reaction buffer: 100 mM potassium-phosphate (pH 8.0), 1.5 mM EDTA, 0.1 mM phenylmethanesulfonyl fluoride (PMSF; add fresh prior use), store at 4 °C.
4. NAD^+: prepare 100 and 50 mM stock solutions (in distilled water), aliquot, and store at −20 °C.
5. α-Lipoamide: prepare 120 mM stock solution (in ethanol), aliquot, and store at −20 °C.
6. NADH: prepare 20 mM stock solution (in distilled water), aliquot, and store at −20 °C.

2.4 Redox-Dependent Modifications of Recombinant mtLPD1

1. Equipment and solutions for SDS-PAGE (as given in Subheading 2.2.4; omitting all reducing chemicals).
2. Spectrophotometer with selectable wavelength (412 nm) and semi-micro plastic cuvettes, preferentially with multiple cuvette holder.
3. Pipettes, 1.5 mL microfuge tubes, and refrigerated microcentrifuge.
4. Purified and desalted recombinant mtLPD1 (as given in Subheading 2.2.2).
5. Protein quantification reagents (as given in Subheading 2.2.3).
6. Ellman's reagent: 2 mM 5,5′-dithiobis-(2-nitrobenzoic acid) (DTNB) in 50 mM sodium acetate, prepare fresh and keep at 4 °C.
7. Desalting columns (polypropylene desalting columns, with a 7 K molecular weight cutoff).
8. Dithiothreitol (DTT): prepare 1 M stock solution (in 1.5 M Tris–HCl, pH 8.8), prepare fresh.
9. Diamide/H_2O_2: 100 mM stock solution (in 1.5 M Tris–HCl, pH 8.8), prepare fresh.
10. Iodoacetamide: prepare 1 M stock solution (in distilled water), aliquot, and store at −20 °C (*see* **Note 7**).

2.5 In Planta Redox State of mtLPD1

2.5.1 Sampling, Protein Precipitation, and MM $(PEG)_{24}$-Alkylation

1. Plant growth chambers, capable for high light (up to 1000 μmol m^{-2} s^{-1} light intensity) and elevated CO_2 (up to 10,000 ppm) experiments if photorespiratory mutants are being evaluated.
2. Leaf material from plants (e.g., wild-type and selected T-DNA insertional mutant plants) grown under standard growth conditions (~150 μmol m^{-2} s^{-1} light intensity, 12/12 h day/night cycle, 22/20 °C day/night temperature, 400 ppm CO_2, 21% O_2, ~70% relative humidity) to growth stage 5.1 [14] and following exposure to an environmental change of interest (shifts to high light or altered CO_2 concentrations) with otherwise equal conditions.
3. Pipettes, 1.5 mL microfuge tubes, and refrigerated microcentrifuge.
4. Liquid nitrogen and suitable containers for storage and transport, mortar and pestle (*see* **Note 8**).
5. Heating blocks for microfuge tubes.
6. 10% (v/v) trichloroacetic acid (TCA) solution, acetone.
7. Alkylation buffer: 50 mM Tris–HCl (pH 7.8); 10 mM (MM $(PEG)_{24}$) methyl-PEG-maleimide alkylation reagent (e.g., Thermo Fisher Scientific: catalog number 22713), 2% (v/v) SDS; 2.5% (v/v) glycerol; 4 M urea, prepare fresh (*see* **Note 9**).
8. Pierce BCA protein assay kit (*see* **Note 10**).

2.5.2 Non-reducing SDS-PAGE, Immunoblotting, and Imaging

1. Equipment and solutions for SDS-PAGE (as given in Subheading 2.2.4; omitting all reducing chemicals).
2. Semidry electroblotting device.
3. Polyvinylidene difluoride (PVDF) membrane.
4. Whatman gel-blotting paper.
5. Methanol (≥99.9%).
6. Transfer-buffer: 1× transfer-buffer, 20% (v/v) methanol, prepare a 10× transfer-buffer stock solution (480 mM Tris–HCl, 390 mM glycine, 0.37% (v/v) SDS).
7. 1× TBS (tris-buffered saline) buffer, prepare a 10× TBS-buffer stock solution, 200 mM Tris–HCl (pH 7.6), 1.4 M NaCl
8. Skimmed milk powder.
9. Antibody specific for mtLPD1 [17] and horse radish peroxidase (HRP)-linked secondary antibody (Goat anti-Rabbit IgG (H + L) Secondary Antibody, HRP).
10. Electrogenerated chemiluminescence (ECL) development kit.
11. Hybridization device with temperature control.
12. Imaging device, including software for image processing and analysis.

3 Methods

All procedures are carried out at room temperature unless otherwise stated. Handling of genetically modified organisms (GMOs) should be carried out in laboratories that are at least S1 security level, essentially following other routine lab security rules. Verified plasmids carrying the gene of interest and sedimented *E. coli* cells following protein overexpression can be stored at −20 °C until further processing. Leaf-material of Arabidopsis plants should be harvested from plants grown under environmentally controlled conditions to growth stage 5.1 [14] and/or following exposure to an environmental change of interest (e.g., CO_2 concentration or high light shift) with otherwise equal conditions. Plant material and extracted and redox labeled proteins should be stored at −80 °C until further processing if necessary.

3.1 PCR-Amplification and Cloning of mtLPD1

This section describes the amplification of mtLPD1 from plant cDNA via PCR [18], using gene-specific oligonucleotides (primers). The oligonucleotides used contain specific restriction enzyme recognition sites for subsequent in frame insertion of the fragment into a protein expression vector. Sequence confirmation of the obtained fragment following insertion into a cloning vector is mandatory to ensure no mutations were introduced during the plasmid generation.

3.1.1 3PCR-Amplification of mtLPD1

1. Amplification of the mtLPD1 coding sequence (CDS) of interest from plant cDNA using a proof-reading DNA polymerase. Pipette a standard 25 μL PCR mixture per reaction in a PCR tube:
   (a) 13.25 μL PCR grade water
   (b) 5.00 μL 5× Q5 buffer
   (c) 2.50 μL forward primer (stock solution 100 μM; 1:20 diluted)
   (d) 2.50 μL reverse primer (stock solution 100 μM; 1:20 diluted)
   (e) 0.50 μL dNTP (stock solution 10 mM)
   (f) 0.25 μL Q5 DNA-polymerase (stock solution 2000 units mL^{-1})
   (g) 1.00 μL template cDNA

2. Run a PCR program with optimized amplification time of the expected fragment size and temperatures for oligonucleotide binding according to in silico prediction.

	Temperature	Duration	Step information
(a)	98 °C	30 s	Initial denaturation
(b)	98 °C	10 s	Cycle (25–35)
(c)	50–72 °C	30 s	Cycle (25–35)
(d)	72 °C	30 s/1 kb	Cycle (25–35)
(e)	72 °C	2 min	Final extension
(f)	20 °C	∞	Infinity hold

3.1.2 Gel Electrophoresis

1. Add 2 μL of DNA loading dye to each completed PCR.
2. Prepare agarose-gel by dissolving and boiling 1 g of agarose (1%) in 100 mL of 1× TAE-buffer (*see* **Note 11**), cool down to 50 °C, add 2.5 μL ethidium bromide and pour unsolidified gel into the gel-chamber.
3. Following gel polymerization, load and separate the PCR amplicons by electrophoresis (80 V, 30 min) and compare the size of the resulting fragments to an appropriate DNA size standard.
4. Obtain and purify fragments of the correct size through a gel-extraction kit and store them at −20 °C until further processing. Measure DNA concentration through spectrophotometric analysis according to standard procedures.

3.1.3 Sub-cloning and Sequence Verification of mtLPD1

1. Ligate the PCR amplified and gel-purified mtLPD1 fragment into the pJET1.2 easy cloning vector. Pipette a 10 μL standard reaction mixture as follows:
   (a) 5.00 μL 2× ligase reaction buffer
   (b) 0.50 μL PEG 4000
   (c) 0.50 μL T4-DNA ligase (stock solution 5 units μL^{-1})
   (d) 0.50 μL pJET1.2 vector (stock solution 50 ng μL^{-1})
   (e) 1.50 μL cleaned PCR product (*see* **Note 12**)
   (f) 2.00 μL PCR grade water
2. Incubate at room temperature (~20 °C) for 5–10 min (*see* **Note 13**).
3. Inactivate enzymatic reaction by heat-treating for 10 min at 65 °C.

4. Transform ligations into a suitable *E. coli* host following standard transformation protocols and grow bacteria on LB-agar plates containing vector-specific antibiotics (amp, final concentration 200 μg mL^{-1}) at 37 °C overnight (~18–24 h).
5. Verify recombinant *E. coli* clones following standard PCR procedures using vector-specific oligonucleotides for amplification and gel electrophoresis.
6. Inoculate and grow putative positive recombinant clones in liquid cultures (e.g., 5 mL LB-medium, including the respective antibiotic). Shake at 37 °C overnight (~18–24 h, 180 rpm).
7. Obtain plasmid DNA from *E. coli* cultures through plasmid DNA isolation following the manufactures protocol.
8. Verify recombinant plasmids through restriction analysis. Pipette a standard 20 μL reaction mixture per isolated plasmid in a 1.5 mL reaction tube:
   (a) 14.60 μL PCR grade water
   (b) 0.20 μL restriction enzyme A (as specified through the forward primer sequence)
   (c) 0.20 μL restriction enzyme B (as specified through the reverse primer sequence)
   (d) 2.00 μL 10× reaction buffer (suitable for both restriction endonucleases)
   (e) 3.00 μL Plasmid DNA [~1–2 μg]
9. Incubate the reaction at 37 °C for 1 h (*see* **Note 14**).
10. Perform gel electrophoresis to separate the fragments as described in Subheading 3.1.2. Include a non-digested control of each isolated plasmid for comparison.
11. Sequence at least two positive recombinant clones following the specifications given by the sequencing company or service.
12. Keep aliquots of the verified plasmids at −20 °C for long-term storage.

3.1.4 Cloning of mtLPD1 for Protein Expression

1. Excise the mtLPD1 fragment from a sequence-positive pJET1.2 clone and cut the protein expression vector (pET28a) with the same restriction enzymes following **step 8** in Subheading 3.1.3.
2. Perform gel-electrophoresis according to Subheading 3.1.2 and gel-elute digested mtLPD1 and pET28a fragments with compatible ends.

3. Pipette the following 20 μL standard reaction mixture to ligate mtLPD1 into pET28a:
   (a) 2.00 μL 10× ligase reaction buffer
   (b) 0.50 μL PEG 4000
   (c) 0.50 μL T4-DNA ligase (stock solution 5 units μL^{-1})
   (d) 2.00 μL pET28a vector
   (e) 6.00 μL mtLPD1 (*see* **Note 15**)
   (f) 9.00 μL PCR grade water
4. Incubate for at least 1 h at room temperature (*see* **Note 16**).
5. Transform the ligation reaction into a suitable *E. coli* host following standard procedures and carry out bacterial growth on LB-agar plates supplemented with the respective antibiotic (e.g., kanamycin for pET28a) at 37 °C overnight (~18–24 h).
6. Verify selected clones via standard PCR- and restriction analysis as described in Subheading 3.1.3, **steps 5–10**.
7. Keep aliquots of two verified plasmids at −20 °C for long-term storage.
8. Re-transform a verified pET28a:mtLPD1 plasmid into a suitable *E. coli* expression strain (BL21gold) according to standard procedures. Perform growth and verification according to **steps 5** and **6** described above.

3.2 Expression and Purification of Recombinant mtLPD1

This section describes the procedure for standard mtLPD1 protein overexpression using verified *E. coli* clones and how to carry out immobilized metal ion affinity chromatography (IMAC) to purify mtLPD1 (*see* **Note 17**). An example of the SDS-PAGE analysis following the mtLPD1 purification process is exemplarily shown in Fig. 1.

3.2.1 Overexpression of mtLPD1

1. Inoculate a verified pET28a:mtLPD1 clone as starter culture in LB-media (5 mL, supplemented with kan, 100 $\mu g\ mL^{-1}$) and grow at 30 °C overnight (~18–24 h; 180 rpm).
2. Dilute the bacteria 1:100 in 200 mL LB-media (supplemented with kan, 100 $\mu g\ mL^{-1}$) and shake (180 rpm) at 37 °C until OD_{600} has reached ~0.6 (*see* **Note 18**).
3. Collect 1 mL of the culture into a 1.5 mL microfuge tube through centrifugation (microfuge 12,000 rpm, 5 min) and save the pellet as uninduced control sample.
4. To the remaining culture add IPTG (1 mM final concentration) and shake (180 rpm) at 37 °C for 4 h for protein overexpression.

5. Collect 1 mL of the culture into a 1.5 mL microfuge tube through centrifugation (microfuge 12,000 rpm, 5 min) and save the pellet as induced control sample.
6. Harvest cells through centrifugation (10 min, 10,000 *g*, 4 °C) and discard supernatant.
7. Add 5–10 mL homogenization buffer to the pellet and resuspend the cells. Store the re-suspended cells at −20 °C if protein purification does not directly follow.

3.2.2 Affinity Purification of mtLPD1

1. Thaw cells on ice, add lysozyme (1 mg mL^{-1}), and shake (120 rpm) at 4 °C for 30 min.
2. Perform ultrasonic cell disruption: use six cycles of 30 s continuous ultrasonic treatment (70 W) followed by a 2 min break. Cells should be kept on ice during all steps.
3. Transfer the cell suspension into a suitable centrifuge tube and centrifuge (10 min, 20,000 *g*, 4 °C).
4. Collect 20 μL of the supernatant into a 1.5 mL microfuge tube (control-soluble fraction).
5. Collect 20 μL of the pellet (resolubilized using the homogenization buffer) fraction into a 1.5 mL microfuge tube (control sample for the insoluble fraction).
6. Equilibrate prepacked Ni-NTA columns with three column volumes each of first of the equilibration buffer and then the homogenization buffer.

   Apply the supernatant to the column to allow protein binding.
7. Collect 20 μL of the flow through into a 1.5 mL microfuge tube (control sample for unbound protein).
8. Wash column with three column volumes of each of first of the homogenization buffer and then the washing buffer to remove excess or unspecific bound proteins.
9. Collect 20 μL of both washing steps into a 1.5 mL microfuge tube (control sample for—unbound and weak interacting protein).
10. Elute recombinant protein via application of three successive column volumes of elution buffer (elution 1–3) into separate 2 mL microfuge tubes.
11. Collect 20 μL of each elution step into a 1.5 mL microfuge tube (control samples for the presence and purity of the protein in each of the elution fraction).
12. To prepare columns for storage, wash column with three successive column volumes of 20% ethanol and store the columns in 20% ethanol at 4 °C.
13. Keep eluted proteins at 4 °C until SDS-PAGE analysis has been carried out.

3.2.3 Protein Quantification, SDS-PAGE, and Immunoblotting

1. Determine the protein concentration of soluble fractions from the purification process defined in Subheading 3.2.2 according to standard assays [12] (*see* **Note 19**).
2. To run SDS-PAGE, mix 10 μg of each soluble protein fraction with 0.5 volume parts of 3× Laemmli buffer (*see* **Note 20**) and boil for 10 min at 95 °C. Also dissolve a small amount of the pelleted cells and the insoluble fraction in 20 μL 3× Laemmli-buffer and boil for 10 min at 95 °C. Put samples on ice for 2 min and spin down the condensate by centrifugation for 30 s at 3000 rpm in a microfuge.
3. Prepare separating gel by mixing 4 mL acrylamide, 3.35 mL distilled water, 2.5 mL 1.5 M Tris–HCl (pH 8.8), 100 μL 10% SDS solution, 50 μL 10% ammonium persulfate, and 5 μL TEMED in a 20 mL Erlenmeyer flask (amounts sufficient for two gels, the first gel is used for a Coomassie stain, serving as loading control, the second gel will be used for the immunoblot) and cast gel within a 10 cm × 10 cm × 4 mm gel cassette. Leave 1–2 cm space in the cassette to allow room for the stacking gel and overlay with water. Wait for complete polymerization of the resolving gel prior to pouring the stacking gel.
4. Prepare stacking gel by mixing 0.975 mL acrylamide, 2.725 mL distilled water, 1.25 mL 0.5 M Tris–HCl (pH 6.8), 50 μL 10% SDS solution, 40 μL 10% ammonium persulfate, and 5 μL TEMED in a 20 mL Erlenmeyer flask (amounts sufficient for two gels, the first gel is used for a Coomassie stain, serving as loading control, the second gel will be used for the immunoblot). Remove the water from the separation gel, fill in the stacking gel, and immediately insert a 12-well gel comb without introducing air bubbles.
5. Load samples, a suitable protein standard and fill 20 μL of 1× Laemmli buffer into each lane to have no empty gel lanes. Ensure to leave out one lane between the protein standard and the protein samples for separating and staining of the size standard (*see* **Note 21**).
6. Run gel at 80 V until samples have entered the separation gel and continue at 140 V until the dye front reaches the bottom of the gel.
7. While the gel is running, prepare two sheets of Whatman paper and one sheet of a PVDF membrane cut to the size appropriate to cover the stacking gel. Saturate Whatman papers with 1× transfer buffer. Saturate the PVDF membrane first with methanol and then flush with 1× transfer buffer.
8. Also, while the gel is running, assemble the semi-dry electroblotting apparatus, wet the anode plate with 1× transfer buffer, and place one sheet of the pre-soaked Whatman paper and the PVDF membrane thereon.

9. Once the gel has finished running, turn off the power supply of the SDS-PAGE, separate the gel plates using a spatula, and remove the stacking gel by cutting it with a fresh razor blade. Rinse the separating gel with water and stain this first gel using Coomassie staining. After 30 min incubation with Coomassie, destain the gel (~3–5 washes with the destaining solution for 10 min shaking each) and fix the stain by applying 7% acetic acid (20 min shaking). Remove the acetic acid and keep the gel in water. Photograph the destained and fixed gel to document the purification process (e.g., *see* Fig. 1).
10. Put the second gel on the membrane for immunoblotting analysis. Ensure to flag the free lane between the protein standard and the protein samples.
11. Place a sheet of Whatman paper on top of the gel, close the blotting device by adding the cathode plate, and switch on the power supply (1 mA cm^{-2} membrane size) for 90 min (*see* **Note 22**).
12. Once the immunoblot has finished running wash the membrane thrice with 1× TBS buffer for 5 min and cut off the marked protein standard lane from the membrane if no pre-stained marker is used. Incubate the standard lane for 10 min in Coomassie solution and destain it as outlined above until the protein standard is visible (*see* **Note 23**).
13. Block the membrane for 1 h in 5% skimmed milk in 1× TBS buffer (shake at room temperature or overnight at 4 °C).
14. Wash the membrane four times in 1× TBS buffer for 5 min each wash. Put the membrane into 5% skimmed milk in 1× TBS buffer, containing an appropriate dilution of the primary antibody, for 1 h (shake at room temperature or overnight at 4 °C) (*see* **Note 24**).
15. Wash the membrane four times in 1× TBS buffer for 5 min each wash. Put the membrane into 5% skimmed milk in 1× TBS-buffer, containing an appropriate dilution of the secondary antibody (*see* **Note 25**), for 1 h (shake at room temperature or overnight at 4 °C).
16. Wash the membrane four times in 1× TBS buffer for 5 min each wash.
17. Develop and image the membrane via the ECL-substrate kit and a suitable imaging device according to the manufacturer's instructions (*see* **Note 26**).

18. If tag-specific affinity purification of the protein resulted in an appropriate purity of the protein, use the elution fraction with the highest protein concentration for enzymatic activity measurements and/or analysis of redox modifications following removal of the affinity tag (thrombin treatment in the case of pET28a) and desalting the elution fractions on PD-10 size exclusion columns (*see* **Note 27**).
19. The surplus of the tag-free and desalted protein can be stored at −80 °C in 20% (v/v) sterile glycerol for other purposes (*see* **Note 28**).

3.3 Enzyme Activity Measurements of mtLPD1

To determine the catalytic parameters of mtLPD1, the enzymatic reaction can be measured in the forward and reverse directions. Here, we provide only the procedure to measure the activity in the forward direction since the procedure for the reverse direction (lipoate reduction) has been described previously [13]. The given example of the experimental setup uses a temperature-controlled spectrophotometer with selectable wavelength, preferentially with multiple cuvette holder to measure multiple reactions.

1. Use the purified, tag-free, and desalted recombinant mtLPD1 protein for biochemical characterization. Store the protein and all substrates on ice unless the activity measurement is carried out.
2. Set spectrophotometer to 340 nm and 25 °C.
3. Pipette the following standard mixture (1 mL) into the cuvette:
   (a) 935 μL reaction buffer
   (b) 30 μL NAD^+ (final concentration 3 mM)
   (c) 10 μL purified mtLPD1 (5–10 μg mL^{-1}) (*see* **Note 29**)
4. Mix and measure baseline for about 2 min until the 340 nm signal is a stable baseline.
5. Once the 340 nm signal is stable, add 25 μL dihydrolipoic acid (DHLA; final concentration 3 mM) to the cuvette to initiate the enzymatic reaction and follow the change in absorbance for 3–6 min (linear phase).
6. Calculate the rate of change in the absorbance of the baseline and after adding the enzyme substrate. Subtract the baseline from the actual measurement and calculate the enzyme activity according to the Lambert–Beer law (molar extinction coefficient of NADH = 6220 M^{-1} cm^{-1}). One unit of enzyme activity is defined as the production of 1 μmol NADH min^{-1} mg $protein^{-1}$.
7. For detailed biochemical characterization, vary the concentration of the co-factor NAD^+ or the enzyme-substrate DHLA, while keeping the other concentration at a saturating level as described elsewhere [10].

3.4 Redox-Dependent Modifications of Recombinant mtLPD1

Cysteine residues of proteins participate in the catalysis of biochemical reactions, are crucial for redox reactions, and influence the protein structure (e.g., as demonstrated for cytosolic malate dehydrogenases [19]). Cysteine thiols (-SH) as reactive residues can be oxidized by reactive oxygen species leading to different redox modifications such as the formation of inter- and intramolecular disulfide bridges (-S-S-) as well as the oxidation of cysteine thiols to sulfenic acid (-SOH) or overoxidation to sulfinic ($-SO_2H$) and sulfonic ($-SO_3H$) acid. To analyze these redox modifications of recombinant mtLPD1, this section describes the procedure for proper redox treatments and how to analyze the redox-dependent modifications visualized via non-reducing SDS-PAGE.

3.4.1 Preincubation of Recombinant mtLPD1

1. Use the elution fraction with the best purified and desalted mtLPD1 protein in 40 mM KPi buffer (pH 8.0) (*see* **Notes 30** and **31**).
2. Add 0.5 mM (final concentration) DTT, mix, and incubate the sample for 30 min at room temperature (Fig. 2a).
3. Remove DTT from the sample using appropriate desalting columns and following the manufacturers protocol.
4. Determine the protein concentration of the desalted sample according to standard assays [12] (Fig. 2a).

3.4.2 Determination of the DTT Concentration After Desalting (Ellman's Test)

1. Prepare a standard DTT calibration curve in a range of 10–500 μM (e.g., 10 μM, 25 μM, 50 μM, 100 μM, and 500 μM).
2. Prepare the Ellman's working reagent by mixing 50 μL of the Ellman's reagent (as described in Subheading 2.2, **item 6**), 100 μL of 1 M Tris–HCl (pH 8.0), and 840 μl water (volume for each sample to be measured) (*see* **Note 32**).
3. Set spectrophotometer to 412 nm.
4. Mix the Ellman's working reagent carefully by gently inverting the reaction tube, pipette the solution into the cuvette, and determine the background.
5. Add 10 μL of the sample to 990 μL Ellman's working reagent, mix well, and incubate at room temperature for 5 min.
6. Measure the optical absorbance at 412 nm for the standard DTT calibration curve and the samples.
7. Create a standard curve by plotting the DTT concentration in μM (x-axis) versus their corresponding absorbance (y-axis).
8. Calculate the sample concentrations using the standard curve (*see* **Note 33**).

3.4.3 Redox Treatment of Recombinant mtLPD1

1. Pipette 15 μg pre-reduced (and desalted) protein in 35 mM Tris–HCl (pH 8.8) and split the sample into three for the following redox treatments a, b, and c (*see* Fig. 2b):
   (a) DTT (reduced).
   (b) Diamide (oxidized).
   (c) Diamide (for subsequent re-reduction).
3. Add 0.5 mM DTT (final concentration) to sample a.
4. Add 0.5 mM Diamide (final concentration) to samples b and c.
5. Mix and incubate the samples for 30 min at room temperature (Fig. 2b).
6. Stop the redox reaction of the DTT-treated (sample a) and one Diamide-treated sample (sample b) by adding 30 mM iodoacetamide (final concentration) (*see* **Note 34**).
7. Add 20 mM DTT (final concentration) to the remaining Diamide-treated sample (sample c; re-reduction) and incubate for 30 min at room temperature (Fig. 2b).
8. Stop the redox reaction of the re-reduced sample (sample c) by adding 30 mM iodoacetamide (final concentration).

3.4.4 Non-reducing SDS-PAGE

1. To run a non-reducing SDS-PAGE mix 5 μg of each redox-treated and alkylated protein sample (as given in Subheading 3.4.2) with 0.5 volume parts of 3× Laemmli-buffer (as given in Subheading 2.2.4) omitting any reductant (e.g., β-mercaptoethanol) and boil for 5 min at 95 °C. Put samples on ice for 2 min and spin down the condensate to collect evaporated solution by centrifugation for 30 s at 3000 rpm.
2. Prepare the separating and stacking gel for SDS-PAGE (as given in Subheading 3.2.3).
3. Load samples, a suitable protein standard and fill similar volumes of 1× Laemmli buffer in every free lane.
4. Run gel at 120 V until the dye front reaches the bottom of the gel.
5. Turn off the power supply of the SDS PAGE, separate the gel plates using a spatula, and remove the stacking gel. Flush the separating gel with water.
6. Stain the gel using the Coomassie staining solution (as given in Subheading 2.2.4) by incubating it at room temperature overnight in the staining solution.
7. Destain the gel using the destaining solution (as given in Subheading 2.2.4) in a horizontal shaker for 1 h at room temperature.
8. Image the destained gel with an appropriate device.

9. To interpret the gel image, compare migration patterns of the reduced (sample a), oxidized (sample b), and re-reduced (sample c) protein. In the case of a redox-regulated protein, reducing conditions (sample a) should lead to the formation of a monomer (e.g., $mtLPD1_{monomer} = 54$ kDa) while oxidizing conditions (sample b) either cause the formation of intermolecular disulfide bridges (a multiple of the monomeric form; e.g., $mtLPD1_{dimer} = 108$ kDa; $mtLPD1_{trimer} = 162$ kDa, etc.) or lead to the formation of intramolecular disulfide bridges (runs lower than the monomeric form) (Fig. 2c). In addition, S-sulfenylation (-SOH), S-sulfinylation (-SO_2H), and/or S-sulfonylation (-SO_3H) can occur under oxidizing conditions that run lower than the monomeric form. Depending on the reversibility of the redox modification, re-reduction (sample c) of a redox-regulated protein should lead to the same result as reducing conditions (sample a) (Fig. 2c). However, since S-sulfinylation (-SO_2H) and S-sulfonylation (-SO_3H) represent irreversible redox modifications due to overoxidation of the cysteine thiols, re-reduction is not possible in this case.

3.5 In Planta *Redox State of mtLPD1*

Given that thiol-based redox regulation is a fundamental posttranslational protein modification to regulate enzymatic activities, it is key to analyze if a specific protein undergoes such modifications in response to environmental changes. This section provides a simplified protocol to determine the *in planta* redox state of mtLPD1 adapted from previously published versions [15, 16]. A schematic overview on the process and potential outcome is summarized in Fig. 3.

3.5.1 Protein Precipitation and MM $(PEG)_{24}$-alkylation

1. Grind the frozen leaf material to a fine powder in liquid nitrogen using a mortar and pestle.
2. Add 10% (v/v) TCA (350 μL per 100 mg leaf-tissue) to quench thiol oxidation.
3. Transfer 100 μL aliquots of the suspension to 1.5 ml microfuge tubes and incubate on ice for 20 min.
4. Centrifuge at 20,000× *g* at 4 °C for 10 min.
5. Wash the pellets with 200 μL acetone (without resuspension) and centrifuge at 20,000× *g* at 4 °C for 10 min.
6. Remove the supernatant and dry the pellets.
7. Resuspend the pellets in 100 μL alkylation buffer and incubate at room temperature for 20 min to allow thiol-alkylation.
8. Heat the samples at 95 °C for 5 min.

9. Centrifuge the samples at room temperature (12,500 rpm, microfuge) to remove residual cellular debris and starch. Transfer the supernatant to a new microfuge tube and determine the protein concentration (Pierce BCA protein assay kit—according to the manufacturers' instruction).
10. Alkylated samples can be stored at −80 °C if SDS-PAGE does not directly follow.

3.5.2 Immunoblotting and Visualization

1. Prepare SDS-gels and SDS-PAGE loading buffer according to the protocol given in Subheading 3.2.3. All reducing reagents (e.g., DTT, ß-mercaptoethanol, etc.) must be omitted.
2. Separate 10 μg protein through non-reducing SDS-PAGE as described in Subheading 3.2.3 using 1 gel for Coomassie staining and 1 for immunoblotting using an mtLPD1-specific antibody.
3. Document the results with an appropriate imaging device, including processing software.
4. For data interpretation, each $MM(PEG)_{24}$ molecule bound to a reduced cysteine (-SH group) adds 1.24 kDa to the total protein mass. Compare migration patterns of mutant or treated protein extracts with those obtained from the wildtype or untreated controls. If a protein does show intermediate reduction/oxidation pattern, calculate the respective ratio thereof (Fig. 3d) (*see* **Note 35**).

4 Notes

1. Prepare a 2.45 μM ethidium bromide stock solution. Use 2.5 μL of the stock solution per 100 mL agarose gel. Ethidium bromide can be exchanged through less hazardous commercially available staining solutions (e.g., SYBR Safe DNA Gel Stain; Invitrogen).
2. The standard conditions and temperature for protein overexpression in *E. coli* is around 3–4 h at 37 °C. However, if protein expression is very low under these conditions, temperature can be reduced (30 or 20 °C) and expression carried out for 10–24 h.
3. Imidazole concentrations for the washing buffer must be determined for each protein to be purified. It is recommended to perform a pilot purification experiment, using different imidazole concentrations (e.g., 20, 40, 60, 80, and 100 mM) or an imidazole gradient on a fast-performance liquid chromatography (FPLC) device.

4. Quantification of the protein concentration can also be carried out through other methods, including commercially available kits (e.g., Pierce BCA protein assay kit; Thermo Fisher Scientific).
5. The mtLPD1 activity is determined in the forward direction in this case. However, the enzymatic activity can also be measured in the reverse direction [13].
6. Activity measurements can also be carried out on a plate reader device following adaptation of the reaction volumes and enzyme amounts.
7. Iodoacetamide is water soluble, but very slowly at room temperature. Therefore, iodoacetamide is recommended to be dissolved in hot water.
8. For a bigger number of samples to be analyzed, leaf material can also be grounded using a ball mill.
9. The $MM(PEG)_{24}$ alkylation reagent can be exchanged by other variants according to previously published protocols [15, 16].
10. The Pierce BCA protein assay kit is used here because of the high SDS content in the buffer that interferes with the Bradford reagent.
11. 1% is used here, but the percentage needs to be adapted according to the fragment sizes expected. The amount of gel needs also to be adapted according to the sample amounts and gel chambers available.
12. The amount of PCR product used for ligation can be increased to 3.5 μL depending on the DNA concentration after the gel-elution step. The DNA concentration should be analyzed through spectrophotometric analysis or gel electrophoresis for optimization.
13. Incubation time can be extended to 20 min for DNA fragments larger than 1500 bp. Longer incubations are not recommended to prevent increased numbers of false positive recombinant clones (according to manufacturer's recommendation).
14. Incubation time for plasmid digestion can be shortened if high-fidelity restriction enzymes are used. Incubation temperature might be adapted according to the requirements of the restriction enzyme.
15. For increased ligation efficiency use a threefold excess of the mtLPD1 fragment compared to the vector. Estimate DNA concentrations through gel electrophoresis for optimization.
16. Prolongation of the incubation time, e.g., overnight (~18–24 h) at 4 °C, might be required to ligate larger DNA fragments.

17. Prior overexpression in larger cultures for protein purification, approximately 4 verified clones should be tested for protein overexpression in smaller scales (5 mL cultures) under standard conditions (3 h, at 37 °C). Expression success should be checked via SDS-PAGE and immunoblotting using a protein specific or tag-specific antibody. If no protein expression can be observed, test expressions should be carried out with lower temperatures and longer expression times. In some cases, omitting the vector-specific antibiotic or using lower concentrations of the inductor of the promoter might be required.
18. In some cases, growth to higher optical densities (OD_{600} ~ 1–1.5) prior induction of overexpression and shorter-expression times might be beneficial to increase protein yield. The volume of the expression culture can be decreased or increased according to the yield in following experiments.
19. The protein concentration should at least be determined from the soluble protein fractions, including the flow through, washing, and elution steps.
20. Separation of 10 μg of soluble protein is recommended for each purification step. In some cases, the amount of protein loaded might be adapted to visualize low abundant contaminations through other proteins.
21. Pre-stained protein size standards can be used to avoid extra treatments of the marker.
22. It is important to remove possible air bubbles between the membrane and the separating gel. Otherwise, the proteins cannot be transferred to the membrane.
23. Do not forget to cut off the protein standard. Once the protein standard is blocked with 5% skimmed milk, Coomassie solution will color the whole membrane stripe.
24. The primary antibody can be specific for the purified protein, if available, or specific for the tag added to the protein (e.g., anti-his-tag).
25. The use of the secondary antibody depends on the source the primary antibody was raised in (rabbit, donkey, goat, etc.).
26. Use a time series for development to ensure proper quantification of the signals.
27. If the protein has not reached a desirable purity, other chromatography steps such as size exclusion or ion-exchange chromatography can be applied. The use of an FPLC system and elution gradients might facilitate the purification process. The elution factions need to be desalted to prevent inhibitory effects during enzyme activity measurements.

28. Storage of recombinant protein in 20% glycerol is recommended but might not always be suitable for the studied protein. Double check for precipitates and measure enzymatic activity after a reasonable time of protein storage.
29. To use an appropriate amount of enzyme, determine the protein concentration of the elution fraction and perform test measurements.
30. The *pKa* value of redox-reactive cysteine thiols is ~8.0. Therefore, buffers should be used with a pH $\geq$ 8.0 (basic conditions) to increase nucleophilicity of the thiol group for redox treatments.
31. The sample volume depends on the protein concentration and the maximum volume capacity of the desalting column. However, the total sample volume at this step needs to be suitable for the following redox treatments and subsequent non-reducing SDS-PAGE (~5 μg protein/sample).
32. Measure each sample and standard as technical duplicates or triplicates.
33. After desalting, the DTT concentration of the sample should be <100 μM to ensure oxidation of the protein by diamide (or other oxidants). The concentration of the oxidant should be ~2.5-fold higher than the reductant left in the sample, otherwise only the reductant might be oxidized but not the protein.
34. Iodoacetamide is a sulfhydryl-reactive alkylating reagent used to block reduced cysteine residues. Usage of iodoacetamide prevents random disulfide formation during SDS-PAGE which would lead to artificial results.
35. Fully reduced and fully oxidized control samples can be prepared using recombinant and purified mtLPD1 following in vitro reduction and oxidation.

References

1. Douce R, Bourguignon J, Neuburger M, Rébeillé F (2001) The glycine decarboxylase system: a fascinating complex. Trends Plant Sci 6:167–176
2. Reed LJ (2001) A trail of research from lipoic acid to alpha-keto acid dehydrogenase complexes. J Biol Chem 276:38329–38336
3. Mooney BP, Miernyk JA, Randall DD (2002) The complex fate of alpha-ketoacids. Annu Rev Plant Biol 53:357–375
4. Timm S, Wittmiß M, Gamlien S, Ewald R, Florian A, Frank M, Wirtz M, Hell R, Fernie AR, Bauwe H (2015) Mitochondrial dihydrolipoyl dehydrogenase activity shapes photosynthesis and photorespiration of *Arabidopsis thaliana*. Plant Cell 27:1968–1984
5. Timm S, Hagemann M (2020) Photorespiration—how is it regulated and how does it regulate overall plant metabolism? J Exp Bot 71:3955–3965
6. Timm S (2020) The impact of photorespiration on plant primary metabolism through metabolic and redox regulation. Biochem Soc Trans 48:2495–2504
7. Reinholdt O, Schwab S, Zhang Y, Reichheld JP, Fernie AR, Hagemann M, Timm S (2019) Redox-regulation of photorespiration through

mitochondrial thioredoxin o1. Plant Physiol 181:442–457

8. da Fonseca-Pereira P, Souza PVL, Hou LY, Schwab S, Geigenberger P, Nunes-Nesi A, Timm S, Fernie AR, Thormählen I, Araújo WL, Daloso DM (2020) Thioredoxin *h2* contributes to theredox regulation of mitochondrial photorespiratory metabolism. Plant Cell Environ 43:188–208
9. Wang Y, Chen X, Spengler K, Terberger K, Boehm M, Appel J, Barske T, Timm S, Battchikova N, Hagemann M, Gutekunst K (2022) Pyruvate:ferredoxin oxidoreductase and low abundant ferredoxins support aerobic photomixotrophic growth in cyanobacteria. elife 11:e71339
10. Kim Y, Ingram LO, Shanmugam KT (2008) Dihydrolipoamide dehydrogenase mutation alters the NADH sensitivity of pyruvate dehydrogenase complex of Escherichia coli K-12. J Bacteriol 190:3851–3858
11. Oñate-Sánchez L, Vicente-Carbajosa J (2008) DNA-free RNA isolation protocols for *Arabidopsis thaliana*, including seeds and siliques. BMC Res Notes 1:93
12. Bradford MM (1976) A rapid and sensitive method for the quantitation of microgram quantities of protein utilizing the principle of protein-dye binding. Anal Biochem 72:248–254
13. Bauwe H (2017) Measurement of enzyme activities. Methods Mol Biol 1653:31–50
14. Boyes DC, Zayed AM, Ascenzi R, McCaskill AJ, Hoffman NE, Davis KR, Gorlach J (2001) Growth stage-based phenotypic analysis of *Arabidopsis*: a model for high throughput functional genomics in plants. Plant Cell 7:1499–1510
15. Suzuki YJ, Marcocci L, Shimomura T, Tatenaka Y, Ohuchi Y, Brelidze TI (2019) Protein redox state monitoring studies of thiol reactivity. Antioxidants (Basel) 22:143
16. Yoshida K, Hisabori T (2019) Simple method to determine protein redox state in *Arabidopsis thaliana*. Bio Protoc 5:e3250
17. Timm S, Florian S, Wittmiß M, Jahnke K, Hagemann M, Fernie AR, Bauwe H (2013) Serine acts as a metabolic signal for the transcriptional control of photorespiration-related genes in *Arabidopsis*. Plant Physiol 162:379–389
18. Mullis K, Faloona F, Scharf S, Saiki R, Horn G, Erlich H (1986) Specific enzymatic amplification of DNA in vitro: the polymerase chain reaction. Cold Spring Harb Symp Quant Biol 51:263–273
19. Liszka A, Schimpf R, Cartuche Zaruma KI, Buhr A, Seidel T, Walter S, Knuesting J, Dreyer A, Dietz KJ, Scheibe R, Selinski J (2020) Three cytosolic NAD-malate dehydrogenase isoforms of *Arabidopsis thaliana*: on the crossroad between energy fluxes and redox signaling. Biochem J 477:3673–3693

Chapter 6

Spectrophotometric Assays for Measuring Hydroxypyruvate Reductase Activity

Aaron H. Liepman, Mara Jaworski, and Citlalxochilt Ramirez-Lopez

Abstract

Hydroxypyruvate reductase (HPR; EC 1.1.1.81) activity is integral to the photorespiratory pathway. Within photorespiration, HPR catalyzes the reduction of hydroxypyruvate, a product of the serine:glyoxylate aminotransferase reaction to glycerate, a substrate for glycerate kinase, using NADH as cofactor. Here we detail a spectrophotometric assay for measuring HPR activity in vitro by following the consumption of NADH at 340 nm.

Key words Photorespiration, Spectrophotometric assay, Hydroxypyruvate reductase, NADH, Enzyme kinetics

1 Introduction

Hydroxypyruvate reductase activity within plant peroxisomes is integral within the photorespiratory pathway. The HPR reaction is shown below:

$$\text{Hydroxypyruvate} + \text{NAD(P)H} + \text{H}^+ \rightarrow \text{D} - \text{glycerate} + \text{NAD(P)}^+$$

Within the photorespiratory pathway, this enzyme converts hydroxypyruvate, a product of the serine:glyoxylate aminotransferase (SGAT; EC 2.6.1.45) reaction, into glycerate, a substrate of glycerate kinase (EC 2.7.1.31). There are at least two HPR isozymes present in plants, including peroxisomal and cytosolic isoforms (*see* **Notes 1** and **2**). These isoforms differ in their activity with various substrates, including hydroxypyruvate and glyoxylate, and co-factors NADH and NADPH [1–4].

HPR activity can be measured in vitro by following the decrease in absorption at 340 nm resulting from oxidation of NADH or NADPH. Assays can be conducted using cuvettes or scaled down in volume for 96-well microplates (*see* **Note 3**), facilitating simultaneous measurements of many samples using a plate-reader UV

Berkley J. Walker (ed.), *Photorespiration: Methods and Protocols*, Methods in Molecular Biology, vol. 2792,
https://doi.org/10.1007/978-1-0716-3802-6_6,

spectrophotometer. This assay can be used for measuring activity of HPR protein purified from plant extracts (e.g., [5, 6]) or purified recombinant HPR protein (e.g., [2, 4]). Here we describe conditions for microplate scaled assays of HPR activity.

2 Materials

Unless otherwise noted, all solutions should be prepared using ultrapure water (18 MΩ × cm at 25 °C). When disposing the waste, appropriate waste disposal guidelines should be followed. Except where noted, stock solutions can be prepared ahead of time, stored frozen (−20 °C), and thawed when needed.

1. Temperature-controlled UV spectrophotometer (standard or plate-reader), set to read absorbance at 340 nm.
2. UV-transparent cuvettes (disposable or quartz) or flat-bottomed 96-well plates (*see* **Note 3**).
3. 1 M potassium phosphate buffer, pH 7.5 (*see* **Note 4**).
4. 1.7 mM NADH, dissolved in 10 mM NaOH (*see* **Note 5**).
5. 20 mM hydroxypyruvate.
6. HPR enzyme dilution buffer: 50 mM Tris–HCl pH 8, 100 mM NaCl, 10% glycerol (*see* **Note 6**).

3 Methods

These activity assays can be conducted using cuvettes (1 mL per sample) or 96-well plates (200 μL per sample). Prepare an assay cocktail of sufficient volume for the desired quantity of samples and replicates. We typically conduct assays in triplicate. When measuring K_m of a substrate, exclude this from the assay cocktail and add it directly to samples to initiate the reaction (*see* **Note 7**). Absorbance data (OD_{340}) are collected in kinetics mode. It is best to use a quantity of enzyme that yields a linear OD_{340} decrease over a 4–6 min period (*see* **Note 8**) and remains linear while approximately doubling in activity when twice the quantity of enzyme is used.

3.1 Hydroxypyruvate Reductase Assay

1. Prepare assay cocktail as indicated in Table 1, mix thoroughly, and store at room temperature (*see* **Note 9**).
2. Dilute enzyme in HPR enzyme dilution buffer. We find that ~15 ng of purified recombinant Arabidopsis HPR1 protein per assay sample yields linear readings throughout the assay duration (Fig. 1) (*see* **Note 10**). Aliquot diluted enzyme into 8-tube PCR tube strips to facilitate multi-channel pipetting.

Table 1
HPR assay components

Reagent and final concentration	Stock concentration	Stock volume per sample (μL)	Volume for ten reactions (96-well format) (μL)
Water	N/A	145	1450
50 mM potassium phosphate pH 7.5	1 M	10	100
0.17 mM NADH	1.7 mM	20	200
500 μM hydroxypyruvate	20 mM	5	50

Reactions are initiated by adding 20 μL of diluted HPR protein, to bring total sample volume to 200 μL

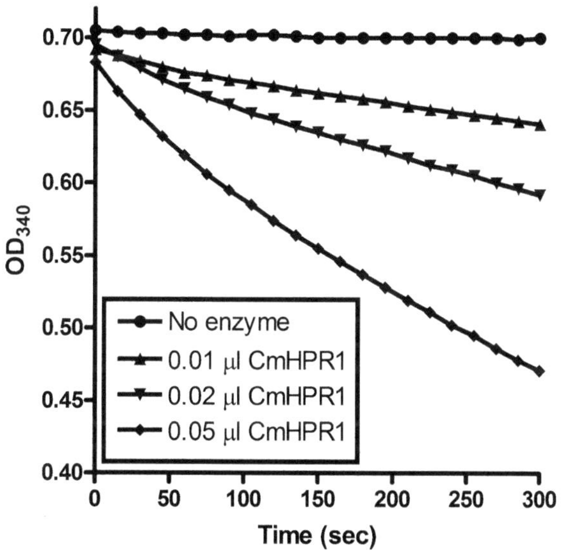

Fig. 1 Representative data from spectrophotometric assays of an HPR protein from *Cyanidioschyzon merolae*. For these assays, various dilutions of purified CmHPR1 protein were tested to determine an appropriate dilution for kinetics assays. 20 μL of diluted enzyme was used for each assay. For example, 20 μL of a 1:2000 dilution is equivalent to 0.01 μL of enzyme stock

3. Set spectrophotometer to measure OD_{340} in kinetics mode at 25 °C (*see* **Notes 11** and **12**), collecting readings at ~15 s intervals.
4. Transfer 180 μL of assay cocktail into designated wells of a 96-well plate.
5. Using an 8-channel pipette, transfer 20 μL of HPR enzyme dilution (or HPR enzyme dilution buffer without enzyme for no enzyme control reactions) from PCR strip tubes into wells of the 96-well plate and mix by pipetting up and down several times. It is important to work quickly at this stage.

6. Immediately initiate kinetics readings. Save/export data to a spreadsheet for calculations.
7. When calculating enzyme activity, use only data within the linear portion of the assay. Subtract the average absorbance change of blank samples (no enzyme controls) from the average absorbance change of samples containing enzyme. Calculate the quantity of NADH consumed per minute using the molar extinction coefficient for NADH at 340 nm (6220 L/M cm) with correction for path length, as necessary (*see* **Note 13**).

4 Notes

1. Plants possess multiple proteins with HPR activity, including peroxisomal and cytosolic isoforms [1–3]. When measuring HPR activity from crude plant cell lysates, total activity will likely include contributions from all isoforms.
2. HPR isoforms differ in their activity with various substrates [1]. For example, Arabidopsis HPR1, the peroxisomal isoform, can reduce hydroxypyruvate using NADH as cofactor. This protein also reacts at a lower rate using glyoxylate and NADPH as substrates. Arabidopsis HPR2, the cytosolic isoform, has similar activity using hydroxypyruvate and glyoxylate as substrates but prefers NADPH to NADH as cofactor [2].
3. We use Corning 9017 96-well plates, though others can be used.
4. 100 mM HEPES-KOH buffer (pH 7.5) also has been used [3].
5. We prepare a fresh NADH stock and use it the same day for assays.
6. Ionic strength of the dilution buffer seems to be important for maintaining activity of diluted enzyme. Purified recombinant Arabidopsis HPR1 rapidly loses activity when diluted in the same buffer with 10 mM NaCl.
7. To determine K_m values for HPR substrates, prepare a modified cocktail lacking the substrate of interest and add the appropriate dilution of the missing substrate directly to samples.
8. We aim for a total OD_{340} change of ~0.1 AU.
9. At final concentrations exceeding 500 μM of hydroxypyruvate, we have observed some inhibition of activity of Arabidopsis HPR1, consistent with previous observations of HPR activity [7] and contrasting with others [4].
10. When assaying purified recombinant AtHPR1, we include ~1.1 U/200 μL assay sample (final concentration of ~5.63 U/mL). This is equivalent to ~15 ng of purified protein per assay.

11. It may be necessary to sample various dilutions of an enzyme preparation to determine an appropriate quantity to achieve assay linearity over a 4–6 min period. It is important to verify that enzyme activity varies proportionally to the quantity of enzyme used for assays. If doubling the amount of enzyme roughly doubles the amount of activity observed, this indicates that an excess of coupler enzyme is present.
12. When conducting assays at other temperatures (e.g., for kinetics measurements at different temperatures), pre-incubate the 96-well plate and assay cocktail at the desired temperature before conducting assays.
13. To calculate enzyme activity for 200 μL assays conducted in 96-well plates, the path length is ~0.571 cm.

Acknowledgment

This work was supported, in part, by a grant from the National Science Foundation (NSF/IOS award #2030295) to A.H.L.

References

1. Givan CV, Kleczkowski LA (1992) The enzymic reduction of glyoxylate and hydroxypyruvate in leaves of higher plants. Plant Physiol 100:552–556. https://doi.org/10.1104/pp.100.2.552
2. Timm S, Nunes-Nesi A, Pärnik T, Morgenthal K, Wienkoop S, Keerberg O, Weckwerth W, Kleczkowski LA, Fernie AR, Bauwe H (2008) A cytosolic pathway for the conversion of hydroxypyruvate to glycerate during photorespiration in *Arabidopsis*. Plant Cell 20:2848–2859. https://doi.org/10.1105/tpc.108.062265
3. Bauwe H (2017) Measurement of enzyme activities. In: Fernie AR, Bauwe H, Weber APM (eds) Photorespiration. Springer, New York, pp 31–50
4. Liu Y, Guérard F, Hodges M, Jossier M (2020) Phosphomimetic T335D mutation of hydroxypyruvate reductase 1 modifies cofactor specificity and impacts Arabidopsis growth in air. Plant Physiol 183:194–205. https://doi.org/10.1104/pp.19.01225
5. Tolbert NE, Yamazaki RK, Oeser A (1970) Localization and properties of hydroxypyruvate and glyoxylate reductases in spinach leaf particles. J Biol Chem 245:5129–5136
6. Titus DE, Hondred D, Becker WM (1983) Purification and characterization of hydroxypyruvate reductase from cucumber cotyledons. Plant Physiol 72:402–408. https://doi.org/10.1104/pp.72.2.402
7. Husic DW, Tolbert NE (1987) NADH:hydroxypyruvate reductase and NADPH:glyoxylate reductase in algae: partial purification and characterization from *Chlamydomonas reinhardtii*. Arch Biochem Biophys 252:396–408. https://doi.org/10.1016/0003-9861(87)90046-4

Chapter 7

High Throughput Glycerate Kinase Activity Assay Using Crude Leaf Extract and Recombinant Enzyme to Determine Kinetic Parameters K_m and V_{max} Using a Microplate Reader

Ludmila V. Roze, Audrey Johnson, and Berkley J. Walker

Abstract

We describe an assay for measuring the activity of D-glycerate 3-kinase (GLYK) in a 96-well microplate format with the use of a set of coupling enzymes. The assay is appropriate for use with a crude protein extract prepared from leaf tissue and with the recombinant purified enzyme. The 96-well microplate format reduces the needed amounts of reagents and coupling enzymes, making the assay less expensive, high throughput, and suitable for the determination of kinetic parameters K_m and V_{max}. In addition, we provide a two-step discontinuous assay modified from past work, making it possible to measure the activity of GLYK at temperatures higher than 45 °C.

Key words Photorespiration, D-Glycerate 3-kinase, Recombinant D-glycerate 3-kinase, 3-Phosphoglycerate

1 Introduction

In chloroplasts, D-glycerate 3-kinase (EC 2.7.1.31) (GLYK) catalyzes the last step in the photorespiratory pathway (PR) to produce 3-phosphoglycerate (3-PGA) from glycerate [1] that enters the C3 cycle thus finishing the recycling of the 2-PG carbon.

Measuring the activity of GLYK involves linking the formation of 3-phosphoglycerate to NADH oxidation with the use of a set of coupling enzymes. The reduced form of NADH absorbs light at 340 nm, and its quantity can be directly monitored by absorbance measured continuously with a spectrophotometer. This method can be used in temperatures ranging from 25 to 45 °C. However, if there is a need to measure activity at a temperature higher than 45 °C, the coupling enzymes denature and become inactive making this continuous method impractical.

In this chapter, we describe an assay for measuring the activity of GLYK in a 96-well microplate format that uses the same

Berkley J. Walker (ed.), *Photorespiration: Methods and Protocols*, Methods in Molecular Biology, vol. 2792,
https://doi.org/10.1007/978-1-0716-3802-6_7,

coupling enzymes as routinely used in rubisco activity assays [2, 3] (Fig. 1). The assay can be used with a crude protein extract prepared from leaf tissue and preparation of a recombinant purified enzyme. The 96-well microplate format reduces the needed amounts of reagents and coupling enzymes, making the assay less expensive, high throughput, and suitable for the determination of kinetic parameters K_m and V_{max}. In addition, we provide a discontinuous two-step assay modified from [3] making it possible to measure the activity of GLYK at temperatures higher than 45 °C.

2 Materials

2.1 Plant Material

The protocol described below provides details for measuring GLYK activity using *Nicotiana tabacum* leaves. However, the GLYK activity measurement could be accomplished with any plant tissue of interest that could be homogenized using a set of standard protocols followed by the optimization of the assay. *Nicotiana tabacum* plants were grown in the greenhouse at 30 °C (90 °F) until tobacco formed five to seven fully developed leaves [4]. Two fully developed top leaves (located next to each other) were used for tissue sampling. For one sample, three punches 17 mm in diameter (A_{total} = 6.81 cm^2, approximately 120 mg) were made, and the punches were placed into 2 mL screw cap plastic tubes, frozen immediately in liquid N_2, and stored at −80 °C. In our laboratory, the GLYK activity measurement using the outlined protocol was performed in the leaf material from *Rhazya stricta* (Rhazya was grown for approx. 1.5 years; six 8.15 mm in diameter punches, A_{total} = 3.14 cm^2, were used to make a homogenate) and birch (at least 1 year old; approximately 120 mg of frozen leaf tissue were used to make a homogenate).

2.2 Common Equipment

1. Cork borer, ~17 mm diameter.
2. 2 mL screw cap microcentrifuge tubes.
3. 1.5 mL snap cap microcentrifuge tubes.
4. Microplate reader.
5. 96-pin microplate replicator (e.g., Boekel Scientific, cat. # 140500).
6. Refrigerated bench-top centrifuges which work with sample sizes of 0.2–2.0 mL and 15–50 mL.
7. Swinging bucket bench-top centrifuge with adapter for 96-well microplates.
8. Pipettes, 2 μL, 20 μL, 300 μL, 1000 μL.
9. Multichannel pipette, 300 μL.
10. Tips for pipettes 10 μL, 200 μL, 300 μL, 1000 μL.

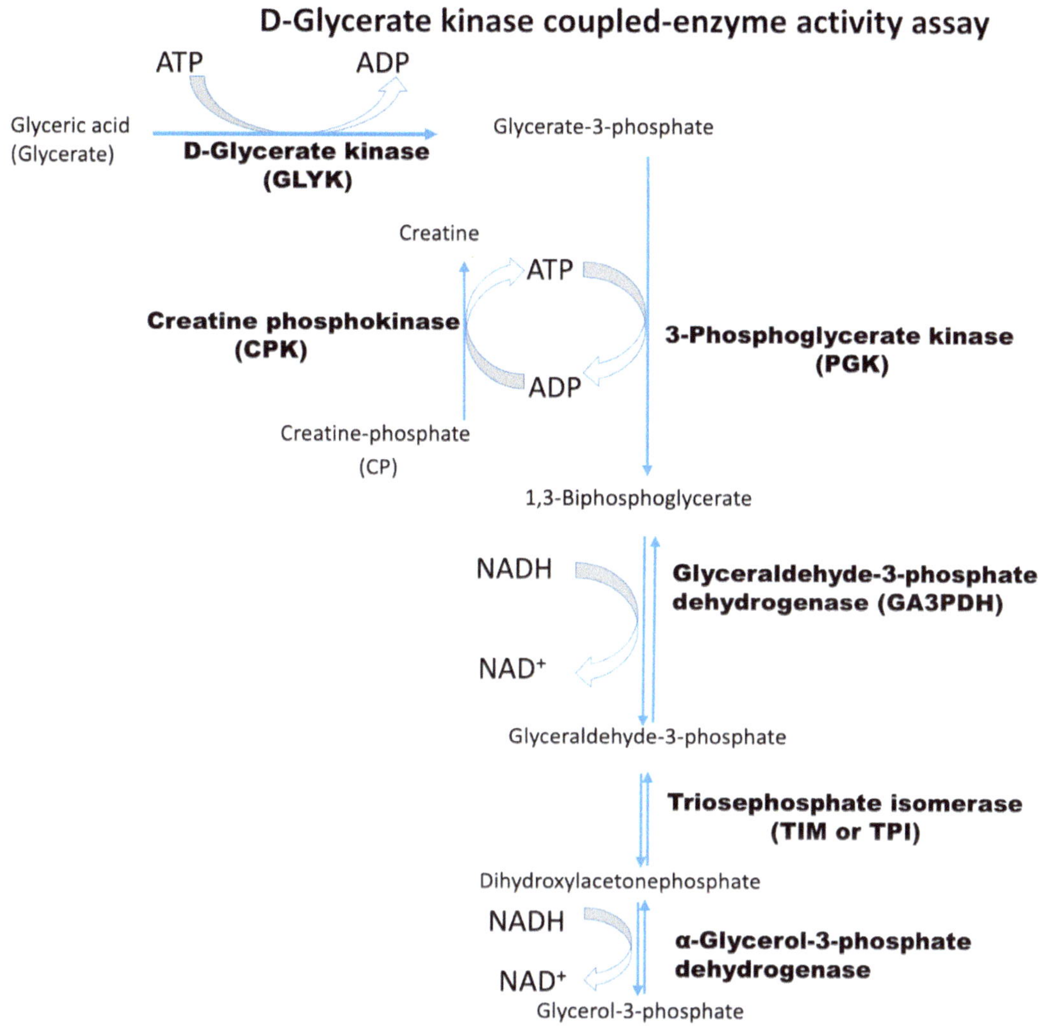

Fig. 1 Schematic of D-glycerate kinase coupled-enzyme activity assay. The assay couples the formation of 3-PGA, the product of glycerate kinase reaction, to 3-phosphoglycerate kinase and ATP to produce 1,3-biphosphoglycerate which is then used sequentially for increased sensitivity by glyceralgehyde-3-phosphate dehydrogenase, triosephosphate isomerase, and α-glycerol-3-phosphate dehydrogenase to produce glycerol-3-phosphate, both dehydrogenases oxidize NADH to NAD^+ resulting in a decrease of absorbance at 340 nm. Creatine phosphokinase serves to regenerate ATP

11. 2 mL glass-to-glass homogenizer (Kontes Glass Co., Vineland, NJ, USA).
12. Reservoir for multichannel pipet.
13. Polystyrene 96-well microplates.
14. Digital Timer.
15. Parafilm.
16. Heating block at 95 °C.

17. Volumetric flask 100 mL.
18. 15 mL and 50 mL plastic tubes.
19. Glass tubes 16 × 125 mm.
20. Autoclavable culture tube caps, plastic, or metal.
21. Dialysis tubing, or dialysis cassette, or other dialysis system.
22. Sterile toothpicks, pointed.
23. Incubator with orbital shaking platform for 37 °C and 20 °C.
24. HisPur™ or other Ni-NTA Spin columns.
25. Sonicator.
26. Benchtop liquid nitrogen container.
27. Biosafety cabinet.
28. Forceps.

2.2.1 GLYK Activity Assay

1. Extraction buffer 50 mM EPPS buffer, pH 8.0, containing 1 mM EDTA, 10 mM DTT, 0.1% Triton X-100 [v/v], 0.5% polyvinylpyrrolidone, and 10 ul 1X Protease Inhibitor Cocktail, EDTA Free. Refer to the Buffer calculation app for amounts of chemicals. Prepare extraction buffer fresh before use.
2. Reaction buffer 50 mM HEPES, pH 7.8, containing 10 mM $MgCl_2$, 60 mM KCl, 5 mM ATP, 5 mM creatine phosphate, and 0.2 mM NADH [5]. Prepare extraction buffer fresh before use. Refer to the buffer calculation app for the amounts of chemicals.
3. Preparation of coupling enzymes (22.5 U mL^{-1} 3-phosphoglycerate kinase, 12.5 U mL^{-1} creatine phosphokinase, 20 U mL^{-1} glyceraldehyde-3-phosphate dehydrogenase, 20 U mL^{-1} glycerol-3-phosphate dehydrogenase, and 56 U mL^{-1} triose-phosphate isomerase):
   (i) Make 50 mM EPPS-NaOH pH 7.8: 1.62 g EPPS in 90 mL. Adjust pH with 25% NaOH, adjust volume to 100 mL. Filter through 0.22 μm filter using a syringe, store +4 °C.
   (ii) Combine the following $(NH_4)_2SO_4$ enzyme suspensions in a 15 mL plastic tube:
   10,000 units 3-phosphoglycerate kinase
   8000 units α-glycerophosphate dehydrogenase-triosephosphate isomerase
   22,600 units triosephosphate isomerase (additional quantities required)
   (iii) Centrifuge at 10,000 rpm, 20 min, + 4 °C.

(iv) Remove supernatant and resuspend the pellet in 6.4 mL 50 mM EPPS-NaOH pH 7.8.

(v) Add the remaining enzymes:

10,000 units creatine phosphokinase

10,000 units glyceraldehyde 3-P dehydrogenase.

(vi) Dialyze overnight against 50 mM EPPS-NaOH, pH 7.8 at +4 °C. To minimize protein precipitation during dialysis, you may dialyze for 1 h against 2 L buffer, then overnight against fresh 2 L buffer.

(vii) Centrifuge at 9500 *g*, +4 °C. Retain the supernatant and measure volume. Add glycerol to 20% final concentration. Adjust volume to 8 mL.

(viii) Dispense 200 mL aliquots. Store at −80 °C.

4. D-glyceric acid sodium salt, 500 mM stock: dilute 50 mg in 943 μL H_2O, aliquot by 100 μL, store at −80 °C.
5. Reaction buffer 1 for discontinuous assay: 50 mM HEPES, pH 7.5, 10 mM $MgCl_2$, 60 mM KCl, 5 mM ATP.
6. Reaction buffer 2 for discontinuous assay: 50 mM HEPES, pH 7.5, 10 mM $MgCl_2$, 60 mM KCl, 15 mM NADH, 2 mM ATP, 10 mM creatine phosphate.

2.2.2 Expression and Purification of Recombinant AtGLYK

1. Kanamycin 50 mg/mL stock, aliquot, and store at −20 °C.
2. Luria–Bertani (LB) medium, liquid.
3. LB agar medium (15 g/L) in Petri dishes containing 50 ug/mL kanamycin.
4. 80% glycerol, sterile.
5. 1 M IPTG (isopropyl ß-D-1-thiogalactopyranoside), filter through a sterile 0.22 μm filter using a syringe, aliquot, store at −20 °C.
6. 50 mM Tris–HCl, 300 mM NaCl, pH 8.0. You may prepare the buffer a day before and store at +4 °C.
7. Lysis buffer: 50 mM Tris–HCl, pH 8.0, 300 mM NaCl, 1 mM EDTA, 0.5 mg/mL lysozyme, 0.1 mg/mL DNAse1, 10 mM $CaCl_2$, proteinase inhibitors, 10 mM imidazole, 5% glycerol. Prepare lysis buffer fresh before use.
8. Buffers containing imidazole for binding, washing and elution prepare fresh, add imidazole to buffers just before use, and keep on ice. Binding buffer 50 mM Tris–HCl, pH 8.0, 300 mM NaCl, 10 mM imidazole. Washing buffer 50 mM Tris–HCl, pH 8.0, 300 mM NaCl, 20 mM imidazole. Elution buffer 50 mM Tris–HCl, pH 8.0, 300 mM NaCl, and 500 mM imidazole.

2.2.3 Protein Concentration Determination

1. Bradford 1× Dye Reagent, store at +4 °C. Bring to room temperature an aliquot necessary for the construction of the standard curve and processing the number of leaf extracts for a particular experiment.
2. Bovine serum albumin (BSA) dilution series of standards (2000 ug/mL, 1500 ug/mL, 1000 ug/mL, 750 ug/mL, 500 ug/mL, 250 ug/mL, 125 ug/mL), store at −80 °C.

3 Preparation of Crude Protein Extract

1. For one sample, collect 3 *N. tabacum* punches (each 17 mm in diameter; $A_{total} = 6.81$ cm^2, total weight approx. 120 mg) using fully developed top leaves located next to each other. Using clean forceps quickly place the punches into a 2 mL screw cap plastic tubes, froze immediately in liquid N_2, and store at −80 °C until analysis. Do not allow it to thaw until analysis. Using the described procedure, collect and freeze as many samples as required for the analyses.
2. Homogenize the frozen three-leaf punches manually in 1 mL of the extraction buffer with the use of a 2 mL glass-to-glass homogenizer (Kontes Glass Co., Vineland, NJ, USA) kept on ice, for about 3–5 min until all tissue is ruptured.
3. Transfer the homogenate into a 1.5 mL plastic tube kept on ice and clarify by centrifugation for 10 min at 13,500 *g*, +4 °C. The supernatant represents the crude protein extract and is used for the enzyme assay immediately (*see* **Note 1**).

4 Expression and Purification of a Recombinant *Arabidopsis thaliana* GLYK (AtGLYK)

Expression of recombinant N-terminal 6xHis tagged AtGLYK containing no targeting peptide was performed using NEB T7 express *Escherichia coli* cell line transformed with a plasmid pET28a AtGLYK (DOE JGI Statement). The plasmid is available upon request. Under described conditions, at least a part of the recombinant protein is soluble according to the SDS-PAGE data. The outlined procedure results in 90% or higher purity of the isolated recombinant protein with a yield of around 6–7 mg/mL.

4.1 Expression of Recombinant AtGLYK

1. Inoculate 3 mL of LB containing kanamycin (add 3 μL kanamycin from 50 mg/mL stock) in a glass tube with a single colony from agar plate (LB + Kan) using a sterile toothpick or sterile 10 μL tip; perform these procedures under a sterile biosafety hood. Cover the tube with a culture tube cap. Grow the culture at 37 °C overnight, 225 rpm.

2. The next day, inoculate 50 mL of pre-warmed LB media (kanamycin 50 μg/mL) in 250 mL flask with 0.5 mL of the overnight culture and grow at 37 °C with vigorous shaking until an OD_{600} of 0.6–0.8 is reached (~2.5 h). Check OD_{600} frequently.
3. Place 1 mL sample into 1.5 mL plastic tube immediately before induction, this sample is non-induced culture. Pellet cells 10 min, 3500 *g*, at +4 °C, discard supernatant, freeze the cell pellet on dry ice, and store at −20 °C until analysis by SDS-PAGE.
4. Precool culture to 20 °C (~5–10 min). Add IPTG to a final concentration of 0.1 mM (50 μL 1 M IPTG/50 mL culture) to induce expression.
5. Incubate the culture overnight at 20 °C; the optical density reaches ~2. Collect a second 1 mL sample. This sample is the induced culture; pellet cells (10 min, 3500 *g*, +4 °C), freeze and store the sample at −20 °C for SDS-PAGE analysis.
6. Harvest the cells from 50 mL culture by centrifugation at 4200 *g* for 20 min, +4 °C, in 50 mL plastic tube.
7. Completely remove supernatant and freeze the cells in dry ice or liquid nitrogen. Use the frozen pelleted cells for purification immediately or store the cell pellet at −80 °C until purification procedure.

4.2 Purification of Recombinant AtGLYK

Binding, washing, and elution are basically performed according to the Qiagen Ni-NTA Spin Kit Handbook [6] with modifications outlined below. Perform all procedures at +4 °C.

1. Prepare 50 mM Tris–HCl, 300 mM NaCl, pH 8.0. Dissolve 6.06 g Tris and 17.53 g NaCl per 1 L water. Adjust pH to 8.0 with concentrated HCl. Keep at +4 °C (*see* **Note 2**).
2. Thaw the cell pellet for approx. 15 min on ice and re-suspend the cells in lysis buffer: add 1.5 mL to the cells obtained from 50 mL induced culture (*see* **Note 3**).
3. Incubate combined cell suspension for 1 h on ice with gentle shaking.
4. Keeping the tube with the cell suspension on ice, sonicate 5 × 10 s using a sonicator (e.g., Branson Sonicator) at high frequency, with 10 sec intervals for cooling.
5. Centrifuge 20 min, 19,800 *g*, +4 °C (centrifugation could be performed in 1.5–2 mL plastic tubes).
6. Collect the supernatant (this is the crude soluble protein extract) into 15 mL conical tube, keep it on ice. Take 20 μL supernatant and freeze at −20 °C for SDS-PAGE analysis.

7. Prepare a Ni-NTA column as recommended in the manufacturer's Handbook with modifications: apply 600 μL of 50 mM Tris–HCl, pH 8.0, 300 mM NaCl, 10 mM imidazole to the column and centrifuge the column at 890 *g* for 2 min and +4 °C.
8. Apply 600 μL supernatant onto an equilibrated HisPur Ni-NTA column placed into a 2 mL collection tube. Centrifuge the column with the tube at 50 *g*, 2 min, +4 °C (*see* **Note 2**). Discard the flow-through but collect 20 μL and store at −20 °C until SDS-PAGE analysis.
9. Repeat applying 600 μL supernatant onto the column followed by centrifugation at 50 *g*, 2 min, +4 °C, until all supernatant has been loaded.
10. Wash 2–3 times by applying 600 μL NP1-20 followed by centrifugation at 890 *g* for 2 min, +4 °C. Discard washes, but collect 20 μL, and store at −20 °C until SDS-PAGE analysis.
11. Place the column into a clean 1.5 mL collection tube. Elute the protein twice with 300 μL of NP1-500 followed by centrifugation at 890 *g* for 2 min, +4 °C (*see* **Note 3**).
12. Combined eluates (2 × 300 μL) (You may keep eluates separately) and dialyze twice against 250 mL (50 mM Tris–HCl, pH 8.0, 0.3 M NaCl) at +4 °C, then overnight against 1000 mL 50 mM Tris–HCl, pH 8.0, 0.3 M NaCl, 5% glycerol (*see* **Note 4**).
13. The next day, if protein precipitation is present, remove the precipitated protein by centrifugation at 14,000 rpm, +4 °C for 10 min.
14. Measure protein concentration in the supernatant using NanoDrop and/or Bradford protocol for microplate.
15. Add 80% sterile glycerol to final 10%, aliquot, and store at −80 °C.
16. Using SDS-PAGE (not provided in this chapter), analyze the non-induced and induced cultures, crude soluble protein extract, flow-through, washes, and eluates.

5 Protein Concentration

Protein concentration is determined by the Bradford method [7] using a microplate reader. The assay is based on binding Coomassie Brilliant Blue G-250 dye, a component of Bradford 1× Dye Reagent to protein and formation of the blue color product under acidic conditions with the absorbance maximum at 595 nm. The assay is performed in a standard 250 μL microplate assay format with using BSA in the linear range of 125–1000 μg/mL, as a standard.

1. Bring to room temperature Bradford 1× Dye Reagent and thoroughly mix by inverting the bottle with the Reagent.
2. Bring BSA standards (2000 μg/mL, 1500 μg/mL, 1000 μg/mL, 750 μg/mL, 500 μg/mL, 250 μg/mL, 125 μg/mL, 0 μg/mL) stored at −80 °C, to room temperature.
3. Pipet 5 μL in triplicate of each standard to the wells to create 0.00, 0.625, 1.25, 2.50, 3.75, 5.00, 7.50 μg/well.
4. Pipet 5 μL unknown protein solution in triplicate (it is necessary to make several dilutions in water) into wells of 96-well microplate; include a blank/background well containing 5 ul water.
5. Using a 300 μL multichannel pipette quickly dispense 250 μL of Bradford Dye Reagent into wells, including the blank well, using clean tips for each set of wells, cover the plate with the parafilm, and mix thoroughly in a microplate reader.
6. Incubate the plate at room temperature for 10 min (at least 5 min and no longer than 1 h). Remove the parafilm and record the absorbance at 595 nm.
7. Determine protein concentration using the microplate reader software which automatically plots standard curve and calculates protein concentration in the unknown samples.

6 Pathlength Determination Method for 96-Well Microplate

To accurately estimate the activity of GLYK using a 96-well format, it is necessary to determine pathlength of the measured absorbance as this depends on the volume and the composition of the reaction mix, as a microplate reader measures the absorbance from the top or bottom of the plate. Here we provide a method to determine the pathlength in a standard flat bottom 96-well microplate with a volume of reaction mix of 200 microliters per well using Beer's law.

1. Make fresh 0.15 mM solution of NADH in a 50 mM HEPES buffer, pH 7.5, 60 mM KCl, 10 mM $MgCl_2$, and determine optical density at 340 nm in a quartz cuvette with pathlength $l = 1$ cm, 25 °C. Measure concentration of NADH using Beer's Law: $Abs = \mathcal{E} \times C \times l$ with extinction coefficient for NADH equal $6.22 \times 10^3\ M^{-1} \times cm^{-1}$, and l equal 1 cm.
2. Dispense 196 μL NADH solution into a well of a polystyrene plate, add 4 μL of extraction buffer, measure optical density at 340 nm and, applying Beer's Law: $Abs = \mathcal{E} \times C \times l$, determine l using extinction coefficient for NADH equal $6.22 \times 10^3\ M^{-1} \times cm^{-1}$ and NADH concentration measured earlier. The pathlength was equal to 0.56 cm at 25 °C in our experiments.

7 Glycerate Kinase Activity Assay

The activity of GLYK is measured by linking the formation of the product of the GLYK reaction 3-phosphoglycerate to NADH oxidation using a set of coupling enzymes analogous to the set of coupling enzymes, excluding carbonic anhydrase, used for measuring the rubisco activity (Fig. 1).

7.1 GLYK Activity Assay to Perform at 25–45 °C

1. Dispense 192 μL reaction buffer containing 50 mM HEPES, pH 7.8, 10 mM $MgCl_2$, 60 mM KCl, 5 mM ATP, 5 mM creatine phosphate, and 0.2 mM NADH into wells of 96-well microplate in triplicate according to the number of samples.
2. To the wells add 4 μL crude protein extract or 4 μL purified GLYK (50 ng) and 4 μL coupling enzymes.
3. Take measurements at 340 nm every 10–15 s for 2 min without added substrate to establish the baseline.
4. Initiate the reaction by adding 2 μL 500 mM D-glycerate (5 mM final) with the use of a 96-well microplate replicator.
5. Monitor the decrease in optical density at 340 nm for 5–10 min. The initial linear rate of reaction is used to express the specific activity as umoles NADH utilized per minute per milligram of total protein. Based on the coupling enzyme set used, formation of every mole of 3-phosphoglycerate corresponds to 2 moles of oxidized NADH (Fig. 1).

7.2 The Discontinuous Two-Step GLYK Activity Assay to Perform at a Temperature Higher Than 45 °C

Protocol

(i) The first step is performed in a 2 mL plastic tube at a temperature higher than 45 °C.

1. Dispense 388 μL reaction buffer 1 into 2 mL plastic tubes and warm up to desired temperature in a heating block.
2. Add 4 μL 500 mM D-glyceric acid, mix.
3. Initiate the reaction by adding 8 μL containing 100–200 ng recombinant GLYK and mix and close the tube. Using a timer keep the tube at the desired temperature for 1 min.
4. Stop the reaction by transferring the tube into a heating block at 95 °C for 5 min, then bring it to room temperature.

(ii) The second step is performed in a 96-well microplate at 30 °C:

1. Dispense 96 μL reaction buffer 2 per well to a pre-warmed plate. The final volume of the reaction mix in the first step allows to make three replicates.
2. Add 100 μL per well of the reaction mix obtained in the first step.

3. Observe the initial optical density at 340 nm after 2 min using the microplate reader.
4. Add 4 μL of coupling enzymes per well and monitor optical density at 340 nm after 2 min using the microplate reader. Add coupling enzymes using a multichannel pipette or a 96-pin microplate replicator if many samples are being analyzed.
5. For activity calculation use the difference between the initial values of optical density and optical density obtained after the addition of coupling enzymes; subtract the background optical density values contributed by coupling enzymes only.

8 *Km* and *Vmax* Determination for the Purified Recombinant AtGLYK

1. Prepare D-glycerate solutions of 10 concentrations according to Table 1, including 0 mM/mL, and keep them at room temperature.

Table 1
D-Glycerate dilutions for K_m and V_{max} determination

	For 1 reaction mix		For 10 reaction mixes	
Glycerate in 200 μL reaction mix, mM	Needed 500 mM glycerate, μL	Needed H_2O, μL	Needed 500 mM glycerate, μL	Needed H_2O, μL
5	2	0	40	0
2	0.8	1.2	16	24
1	0.4	1.6	8	32
0.5	0.2	1.8	4	36
	Needed 50 mM glycerate, μL	Needed H_2O, uL		
0.2	0.8	1.2	16	24
0.1	0.4	1.6	8	32
0.05	0.2	1.8	4	36
	Needed 5 mM glycerate, μL	Needed H_2O, μL		
0.02	0.8	1.2	16	24
0.01	0.4	1.6	8	32
0	0	2	0	40

2. Make reaction buffer 50 mM HEPES, pH 7.8, 10 mM $MgCl_2$, 60 mM KCl, 5 mM ATP, 5 mM creatine phosphate, 0.2 mM NADH.
3. Dispense 192 μL reaction buffer into wells of a 96-well microplate in triplicate according to the number of samples.
4. To the wells add 4 μL purified GLYK containing 50 ng enzyme and 4 μL coupling enzymes.
5. Take measurements at 340 nm every 10–15 s for 2 min without added substrate to establish the baseline.
6. Initiate the reaction by addition of 2 μL D-glycerate of different concentrations including 0 mM with the use of a 96-well microplate replicator.
7. Monitor the decrease in optical density at 340 nm for 5 min. Subtract the baseline. The initial linear rate of reaction is used to express the specific activity as μmoles NADH utilized per minute per milligram of total protein. Note that based on the coupling enzyme set used, the formation of every mole of 3-phosphoglycerate corresponds to 2 moles of oxidized NADH.
8. Calculate specific GLYK activity at different concentrations of glycerate.
9. Determine K_m and V_{max} using the R Shiney App available here: (https://github.com/PerennialDr/MichaelisMenten_fit) or any other compatible fitting method.

9 Notes

1. The crude protein extract prepared using tobacco leaves and stored at +4 °C overnight loses approximately 10% of the activity determined immediately after the extract was made.
2. For dialysis prepare 50 mM Tris–HCl, pH 8.0, 300 mM NaCl, fresh or a day before and store at +4 °C.
3. When larger quantities of purified enzyme are required, it is possible to use 3–4 50 mL separate cultures, or simply one culture of larger volume. Careful attention should be paid to aeration during culture growth. For easier handling, the harvested and frozen cells could be combined for lysis and following sonication in one 50 mL tube.
4. Low speed during centrifugation provides more time for the enzyme to bind the Ni-matrix.
5. The elution of recombinant GLYK could be performed by 300 mM imidazole.
6. 300 mM NaCl and 5% glycerol in the dialysis buffer help to prevent protein precipitation.

References

1. Boldt R, Edner C, Kolukisaoglu Ü, Hagemann M, Weckwerth W, Wienkoop S, Morgenthal K, Bauwe H (2005) d-GLYCERATE 3-KINASE, the last unknown enzyme in the photorespiratory cycle in Arabidopsis, belongs to a novel kinase family. Plant Cell 17:2413–2420
2. Kubien DS, Brown CM, Kane HJ (2011) Quantifying the amount and activity of rubisco in leaves. In: Carpentier R (ed) Photosynthesis research protocols. Humana Press, Totowa, pp 349–362
3. Walker BJ, South PF, Ort DR (2016) Physiological evidence for plasticity in glycolate/glycerate transport during photorespiration. Photosynth Res 129:93–103
4. Gregory LM, Roze LV, Walker BJ (2023) Increased activity of core photorespiratory enzymes is associated with higher photosynthetic rates under elevated temperatures in the extremophile *Rhazya stricta*. Plant Cell Environ:1–17. https://doi.org/10.1111/pce.14711
5. Kleczkowski LA, Randall DD (1988) Substrate stereospecificity of leaf glycerate kinase from C3 and C4 plants. Phytochemistry 27:1269–1273
6. Ni-NTA Spin Kit Handbook – (EN) – QIAGEN. https://www.qiagen.com/us/resources/resourcedetail?id=3fc8c76d-6d21-4887-9bf8-f35f78fcc2f2&lang=en. Accessed 10 Mar 2023
7. Quick Start™Bradford Protein Assay

Chapter 8

Purification of Recombinant N-terminal Histidine-Tagged *Arabidopsis thaliana* Phosphoglycolate Phosphatase 1, Glycolate Oxidase 1 and 2, and Hydroxypyruvate Reductase 1

Mathieu Jossier, Céline Oury, Nathalie Glab, and Michael Hodges

Abstract

To measure the kinetic properties of photorespiratory enzymes, it is necessary to work with purified proteins. Protocols to purify photorespiratory enzymes from leaves of various plant species require several time-consuming steps. It is now possible to produce large quantities of recombinant proteins in bacterial cells. They can be rapidly purified as histidine-tagged recombinant proteins by immobilized metal affinity chromatography using Ni^{2+}-NTA-agarose. This chapter describes protocols to purify several *Arabidopsis thaliana* His-tagged recombinant photorespiratory enzymes (phosphoglycolate phosphatase, glycolate oxidase, and hydroxypyruvate reductase) from *Escherichia coli* cell cultures using two bacterial strain-plasmid systems: BL21(DE3)-pET and LMG194-pBAD.

Key words Photorespiration, Phosphoglycolate phosphatase, Glycolate oxidase, Hydroxypyruvate reductase, His-tagged recombinant protein, Immobilized metal affinity chromatography, Rapid one-step protein purification, *Escherichia coli*

1 Introduction

The photorespiratory cycle is made up of eight core enzymes located in three cell compartments (chloroplasts, peroxisomes, and mitochondria). To better understand their kinetic properties and eventual regulatory mechanisms, it is necessary to work with purified proteins. There are examples of protocols to purify photorespiratory enzymes from leaves of various plant species including tobacco [1], spinach [2, 3], Arabidopsis [4, 5], corn [6], and cucumber [7]. However, they are time-consuming and require multiple steps such as ammonium sulfate precipitation, acetone fractionation, overnight dialysis, ion-exchange chromatography, gel filtration chromatography, and occasionally (depending on the enzyme) affinity chromatography steps. For example, the following

Berkley J. Walker (ed.), *Photorespiration: Methods and Protocols*, Methods in Molecular Biology, vol. 2792,
https://doi.org/10.1007/978-1-0716-3802-6_8,

steps were required to purify spinach leaf PGLP to 94% homogeneity: ammonium sulfate precipitation, dialysis, DEAE-Sephacel™ ion-exchange chromatography, ultra-centrifugation, DEAE-cellulose 52 ion-exchange chromatography, and Sephacryl™ S-200 gel filtration chromatography [2].

There are now many systems to produce and rapidly purify large quantities of recombinant proteins. To produce soluble, non-toxic recombinant proteins, the *E. coli* BL21(DE3)-pET expression system is often used. It is based on target genes being under the control of strong bacteriophage T7 transcription. The T7 RNA polymerase gene has been introduced into the *E. coli* BL21(DE3) strain under the control of an IPTG-inducible *lac* operon promoter. When induced, the recombinant protein is produced under the control of a T7 promoter (which is not naturally present in *E. coli*) [8]. This system often leads to high yields of recombinant protein; the desired protein can comprise more than 50% of the total cell protein after only a few hours of induction. To produce poorly soluble proteins, an additional maltose binding protein (MBP) moiety can be added to the recombinant protein to enhance protein solubility and promote proper protein folding [9]. MBP can also be used for amylose affinity-chromatography purification but since this is not very efficient, an additional histidine-tag (His--tag) is often incorporated to facilitate recombinant protein purification. To produce problematic toxic and/or insoluble recombinant proteins, the pBAD-LMG194 bacterial expression system can be used. It is based on the araBAD operon that controls *E. coli* L-arabinose metabolism. The araBAD promoter is induced by the addition of arabinose and thus controls recombinant protein expression in *E. coli* strains lacking L-arabinose catabolism like LMG194. Furthermore, pBAD contains the bacterial AraC gene that fixes to and inhibits the araBAD promoter, in the presence of arabinose this regulatory protein changes conformation, and inhibition is released [10]. Together, these properties allow a tight suppression of basal expression in the absence of arabinose, and recombinant protein expression is more controlled and highly inducible when compared to the BL21(DE3)-pET system, but recombinant protein production is not as high.

Whatever the bacterial expression system used to produce recombinant proteins, a common method is to include a His-tag at either the N-terminal or C-terminal of the protein of interest, produce the His-tagged recombinant protein in *E. coli* cells and then purify the resulting protein using immobilized metal affinity chromatography (IMAC) [11]. This affinity chromatography technique using a nickel-charged resin (Ni^{2+}-NTA-agarose) has proven to be a powerful way to purify His-tagged recombinant proteins. Indeed, histidine has the strongest interaction with transition metals like Ni^{2+}. Nonspecific interactions can be reduced by the addition of 10 mM 2-mercaptoethanol, NaCl (up to 500 mM), and

20% glycerol to wash buffers. His-tagged recombinant proteins are often eluted by the addition of the histidine analogue, imidazole. However, they can also be eluted by lowering the pH or by the addition of EDTA (that removes the Ni^{2+} and the fixed protein from the Ni^{2+}-NTA-agarose column) [11].

In the literature, a number of photorespiratory enzymes have been successfully produced and purified as recombinant proteins. The pBAD-LMG194 system has been used to purify recombinant His-tagged Arabidopsis HPR (hydroxypyruvate reductase) 1 [5, 12] and HPR2 [13], Arabidopsis phosphoglycolate phosphatase 1 (PGLP1) [14] and Synechocystis GDC (glycine decarboxylase complex)-P and GDC-H sub-units [15]. The BL21(DE3)-pET system has enabled the purification of His-tagged Arabidopsis GOX (glycolate oxidase) 1 and GOX2 [16, 17], corn GO1 (glycolate oxidase 1) [16, 17], and Arabidopsis PGLP1 [18]. On the other hand, recombinant Arabidopsis D-glycerate 3-kinase was produced using a different *E. coli* BL21 expression system and purified by Matrex Green A affinity chromatography [4] and although *E. coli* BL21 strain and the pET28a plasmid were used to produce recombinant Arabidopsis glutamate:glyoxylate aminotransferase 1, the resulting protein was purified using various chromatography procedures [19].

In this chapter, we will describe protocols to purify His-tagged recombinant *Arabidopsis thaliana* PGLP1, GOX1, GOX2, and HPR1 using the BL21(DE3)-pET and LMG194-pBAD bacterial systems mentioned above.

2 Materials

2.1 General

1. Icebox with ice.
2. Precision balance.
3. Spatula.
4. pH meter.
5. Magnetic stirrer.
6. Magnetic stirrer bar.
7. Precision pipettes (0–20 μL, 0–200 μL, 0–1000 μL, 0–5000 μL).
8. Glass and plastic beakers (2 L, 1 L, 500 mL, 250 mL).
9. Glass and plastic graduated cylinders (2 L, 1 L, 500 mL 250 mL).
10. Glass bottles (500 mL, 250 mL).
11. Ultrapure water and bidistilled water (*see* **Note 1**).

2.2 Bacterial Culture and Induction of Recombinant Protein Expression

1. Temperature-controlled cell culture shaker (37 °C and 30 °C) set to 1.5× *g* (200 rpm).
2. Laminar flow hood.
3. Spectrophotometer with wavelength set to 600 nm.
4. 50 mL sterile glass tubes.
5. 1 L sterile conical flasks.
6. Sterile inoculation loop.
7. For *At*GOX1/2 and *At*PGLP1 recombinant proteins: Sterile bacterial Luria-Bertani (LB) broth culture medium (*see* **Note 2**).
8. For *At*HPR1 recombinant protein: Sterile bacterial Rudic culture medium (RM) (*see* **Note 2**).
9. For *At*GOX1/2 recombinant protein: 50 g/L stock solution of kanamycin (*see* **Note 3**).
10. For *At*PGLP1 and *At*HPR1 recombinant proteins: 100 g/L stock solution of ampicillin (*see* **Note 3**).
11. For *At*GOX1/2 and *At*PGLP1 recombinant proteins: 1 M stock solution of isopropyl-1-D-thiogalactopyranoside (IPTG) (*see* **Note 4**).
12. For *At*HPR1 recombinant protein: 20% (w/v) L-arabinose (*see* **Note 5**).

2.3 Bacterial Soluble Protein Extraction

1. Refrigerated centrifuge at 4 °C with swinging buckets capable of 1400× *g*.
2. Refrigerated centrifuge at 4 °C capable of 35,000× *g*.
3. French pressure cell press.
4. Pressure cell for French pressure cell press cooled to 4 °C.
5. 250 mL centrifuge bottles cooled to 4 °C.
6. 50 mL conical plastic centrifuge tubes.
7. Aluminum foil.
8. 30 mL centrifuge tubes rated to at least 35,000× *g* and cooled to 4 °C.
9. 0.5 M Tris–HCl, pH 7.5 and 0.5 M Tris–HCl, pH 8 stock solutions (*see* **Note 6**).
10. 3 M NaCl stock solution (*see* **Note 7**).
11. 50% (v/v) glycerol stock solution (*see* **Note 8**).
12. 0.5 M $MgCl_2$ stock solution (*see* **Note 9**).
13. 1 M imidazole stock solution (*see* **Note 10**).
14. 0.5 M stock solutions of Na_2HPO_4 and NaH_2PO_4 for the sodium phosphate buffer (*see* **Note 11**).

15. 10 mM stock solution of flavine mononucleotide (FMN) (*see* **Note 12**).
16. EDTA-free Protease Inhibitor Cocktail Tablets (*see* **Note 13**).
17. *At*GOX1/2 extraction buffer prepared from stock solutions in ultrapure water and cooled at 4 °C: 50 mM Tris–HCl, pH 8, 1 M NaCl, 10% (v/v) glycerol, 10 mM imidazole, 100 μM FMN (*see* **Note 12**), protease inhibitor cocktail (*see* **Note 13**).
18. *At*PGLP1 extraction buffer prepared from stock solutions in ultrapure water and cooled to 4 °C: 50 mM Tris–HCl, pH 7.5, 150 mM NaCl, 10 mM $MgCl_2$, 10 mM imidazole, protease inhibitor cocktail (*see* **Note 13**).
19. *At*HPR1 extraction buffer prepared from stock solutions in ultrapure water and cooled to 4 °C: 50 mM sodium phosphate buffer (pH 8), 300 mM NaCl, 5 mM imidazole, protease inhibitor cocktail (*see* **Notes 11** and **13**).

2.4 Recombinant Protein Purification

2.4.1 General Materials for Batch and Gravity-Flow Affinity Chromatography

1. Rolling wheel at 4 °C (*see* **Note 14**).
2. Immobilized metal ion affinity chromatography (IMAC) resin used for the purification of His-tagged proteins (*see* **Note 15**).
3. 50 mL conical plastic centrifuge tubes.
4. 15 mL conical plastic centrifuge tubes.
5. Vacuum pump.
6. *At*GOX1/2 washing buffer prepared from stock solutions in ultrapure water and cooled to 4 °C: 50 mM Tris–HCl, pH 8, 1 M NaCl, 10% (w/v) glycerol, 20 mM imidazole, 100 μM FMN (*see* **Note 12**).
7. *At*PGLP1 washing buffer prepared from stock solutions in ultrapure water and cooled to 4 °C: 50 mM Tris–HCl, pH 7.5, 150 mM NaCl, 10 mM $MgCl_2$, either 20 mM imidazole when using pDEST-His_6MBP-PGLP1 or 75 mM imidazole when using pET16b-PGLP1.
8. *At*HPR1 washing buffer prepared from stock solutions in ultrapure water and cooled to 4 °C: 50 mM sodium phosphate buffer, pH 8, 300 mM NaCl, 20 mM imidazole (*see* **Note 11**).
9. *At*GOX1/2 elution buffer prepared from stock solutions in ultrapure water and cooled to 4 °C: 50 mM Tris–HCl, pH 8, 1 M NaCl, 10% (w/v) glycerol, 250 mM imidazole, 100 μM FMN (*see* **Note 12**).
10. *At*PGLP1 elution buffer prepared from stock solutions in ultrapure water and cooled to 4 °C: 50 mM Tris–HCl, pH 7.5, 150 mM NaCl, 10 mM $MgCl_2$, 250 mM imidazole.
11. AtHPR1 elution buffer prepared from stock solutions in ultrapure water and cooled to 4 °C: 50 mM sodium phosphate buffer, pH 8, 300 mM NaCl, 250 mM imidazole (*see* **Note 11**).

2.4.2 Additional Materials for Gravity-Flow Affinity Chromatography Purification

1. Glass wool.
2. 5 mL syringe.
3. Clamp stand.

2.5 Recombinant Protein Desalting

2.5.1 Recombinant Protein Desalting by Gravity Flow

1. PD-10 desalting columns (*see* **Note 16**).
2. Clamp stand.
3. *At*GOX1/2 desalting buffer prepared from stock solutions in ultrapure water and cooled to 4 °C: 50 mM Tris–HCl, pH 8, 20% (w/v) glycerol, 100 μM FMN (*see* **Note 12**).
4. *At*HPR1 desalting buffer prepared from stock solutions in ultrapure water and cooled to 4 °C: 50 mM sodium phosphate buffer, pH 8 (*see* **Note 11**).

2.5.2 Additional Materials for Recombinant Protein Desalting by Centrifugation

1. Refrigerated centrifuge cooled to 4 °C with swinging bucket capable of 1000× *g*.
2. *At*PGLP1 desalting buffer prepared from stock solutions in ultrapure water and cooled to 4 °C: 50 mM Tris–HCl, pH 7.5, 10 mM $MgCl_2$.

2.6 Recombinant Protein Quantification

1. Spectrophotometer with wavelength set to 595 nm.
2. Commercial Bradford reagent (*see* **Note 17**).
3. Bovine serum albumin (BSA) for calibration curve for protein quantification by the Bradford method [20].

3 Methods

3.1 Bacterial Culture and Recombinant Protein Production

All following steps must be done under a laminar flow hood to maintain sterile conditions.

3.1.1 His_6-AtGOX1/2 Bacterial Culture and Recombinant Protein Production

1. Take three colonies of freshly transformed *E. coli* BL21(DE3) by pET28a-GOX1/2 plasmid [15] with a sterile inoculation loop and resuspend in a 50 mL glass tube containing 3 mL of LB broth medium and add 3 μL of kanamycin stock solution (50 mg/L final). Incubate overnight at 37 °C in a cell culture shaker set at 200 rpm.
2. Transfer the 3 mL overnight culture by pouring it into a 1 L conical flask containing 200 mL of LB medium (*see* **Note 18**) and add 200 μL of kanamycin stock solution (50 mg/L final) and incubate at 37 °C in a cell culture shaker until an OD_{600nm} of 0.4–0.6.

3. Add 200 μL of 1 M IPTG stock solution (1 mM final) to induce recombinant protein expression and further incubate for 20 h at 30 °C in a cell culture shaker.

3.1.2 His_{10}-AtPGLP1 and His_6-MBP-AtPGLP1 Bacterial Culture and Recombinant Protein Production

1. Take three colonies of freshly transformed *E. coli* BL21(DE3) by pET16b-*At*PGLP1 or pDEST-His_6MBP-PGLP1 plasmid [18] with a sterile inoculation loop and resuspend in a 50 mL glass tube containing 3 mL of LB broth medium and add 3 μL of ampicillin stock solution (100 mg/L final). Incubate overnight at 37 °C in a cell culture shaker.
2. Transfer the 3 mL overnight culture by pouring it into a 1 L conical flask containing 200 mL of LB medium (*see* **Note 18**) and add 200 μL of ampicillin stock solution (100 mg/L final) and incubate at 37 °C in a cell culture shaker until an OD_{600nm} of 0.8 (*see* **Note 19**).
3. Add 100 μL of 1 M IPTG stock solution (500 μM final) to induce recombinant protein expression and further incubate for 4 h at 37 °C in a cell culture shaker.

3.1.3 His_6-AtHPR1 Bacterial Culture and Recombinant Protein Production

1. Take three colonies of freshly transformed *E. coli* LMG194 by pBAD/His-*At*HPR1 plasmid [5, 12] with a sterile inoculation loop and resuspend in a 50 mL glass tube containing 10 mL of RM medium and add 2.5 μL of ampicillin stock solution (50 mg/L final). Incubate overnight at 37 °C in a cell culture shaker.
2. Transfer the 10 mL overnight culture by pouring it into a 1 L conical flask containing 200 mL of RM medium and add 100 μL of ampicillin stock solution (50 mg/L final) and incubate at 37 °C in a cell culture shaker until an OD_{600nm} of 0.7.
3. Add 200 μL of 20% (w/v) L-arabinose (0.2% (w/v) final) to induce recombinant protein expression and incubate for 20 h at 37 °C in a cell culture shaker.

3.2 Preparation of Bacterial Soluble Protein Extracts

1a. Transfer the 200 mL bacterial culture to a 250 mL centrifuge bottle and pellet the cells by centrifugation at 3300× *g* for 30 min at 4 °C.

1b. Alternatively, transfer half of the 200 mL bacterial culture to 2 × 50 mL centrifuge tubes and centrifuge at 3300× *g* for 30 min at 4 °C. After supernatant elimination, transfer the remaining 100 mL of the culture to the same 2 × 50 mL tubes and centrifuge at 1400× *g* for 30 min at 4 °C. Discard the supernatant (*see* **Note 20**).

All of the following steps must be done on ice and protected from the light as much as possible when purifying *At*GOX1/2.

2. Gently resuspend each obtained bacterial cell pellet in 3 mL of the appropriate extraction buffer according to the recombinant protein with a 5 mL precision pipette (*see* **Note 21**). Transfer the resuspended pellet into a 50 mL tube (*see* **Note 12**).
3. Transfer the 3 mL of resuspended bacteria into the pressure cell using a 5 mL precision pipette and place the cell onto the French press and break the bacteria by three passages through the French pressure cell press at 500–600 Pa (*see* **Note 22**). During the final passage, collect the broken bacterial extract in a 30 mL centrifuge tube, increase the volume by adding 5 mL of the appropriate extraction buffer according to the recombinant protein, and gently homogenize.
4. Centrifuge the broken bacterial extract at 35,000× *g* for 30 min at 4 °C. Keep it on ice prior to recombinant protein purification either by gravity flow affinity chromatography (Subheading 3.3.1, **step 2**) or "in batch" affinity purification (Subheading 3.3.2, **step 2**).

3.3 Recombinant Protein Purification

3.3.1 Recombinant Protein Purification by Gravity Flow Affinity Chromatography for His_6-*AtGOX1/2* and His_6-*AtHPR1*

1. Centrifuge 2 mL of Ni^{2+}-NTA-agarose IMAC resin (*see* **Note 23**) in a 50 mL centrifuge tube at 5400× *g* for 10 min at 4 °C. Discard the supernatant by aspiration with a vacuum pump and wash the resin by resuspending in 40 mL of the appropriate extraction buffer according to the recombinant protein to be purified. Centrifuge again at 5400× *g* for 10 min at 4 °C, aspirate the supernatant, resuspend the resin in 10 mL of the appropriate extraction buffer (*see* **Note 24**), and transfer to a 15 mL centrifuge tube.
2. Once the centrifugation step (in Subheading 3.2, **step 4**) has finished, transfer the supernatant from the cell extraction (*see* **Note 25**) to the 15 mL centrifuge tube containing the washed IMAC resin. Incubate on a rolling wheel for 2 h at 4 °C (*see* **Note 14**).
3. At the bottom of a 5 mL plastic syringe (without a needle), place a homogeneous layer of glass wool compacted with the plunger of the syringe. The glass wool layer must have a thickness of about 2 mm once compacted. Fix the syringe to a clamp stand (*see* **Note 12**). Pour the mix of resin and bacterial soluble protein extract into the syringe and let the solution flow by gravity, the resin is retained by the glass wool to form the affinity chromatography column.
4. Carefully and gradually pipette 40 mL of the appropriate washing buffer into the syringe to wash the resin so as to remove proteins that do not interact with the Ni^{2+}. You can discard this flow through liquid.
5. Carefully add 500 μL of the appropriate elution buffer to the top of the resin and let it flow through the resin. Discard this

initial elution fraction as it does not yet contain recombinant protein.

6. Carefully add 2.5 mL of the appropriate elution buffer to the top of the resin and collect the eluate-containing recombinant protein in a 15 mL tube kept on ice (see **Note 12**).

3.3.2 "Batch" Affinity Purification of Recombinant Proteins for His_6-At*GOX1*, His_{10}-At*PGLP1*, and His_6-MBP-At*PGLP1*

1. Add 2.5 mL of Ni^{2+}-NTA-agarose IMAC resin (*see* **Note 23**) to a 15 mL centrifuge tube and add 10 mL of the appropriate extraction buffer and centrifuge in a swinging bucket at 5400× *g* for 6 min at 4 °C. Discard the supernatant by vacuum aspiration and wash the resin two more times: Add 12 mL of the appropriate extraction buffer and carefully resuspend the resin after each washing step (*see* **Note 24**). After the final wash step centrifugation, carefully remove by vacuum aspiration the supernatant but keep 1 mL of extraction buffer on top of the resin. Keep the tube containing the resin on ice.
2. Once the centrifuge step in Subheading 3.2, **step 4** is finished, transfer the supernatant containing the bacterial soluble proteins (*see* **Note 25**) to the 15 mL centrifuge tube containing the washed resin and fill to 13 mL with the appropriate extraction buffer (*see* **Note 12**). Carefully mix the suspension (*see* **Note 24**) and incubate at 4 °C on a rolling wheel for 2 h (*see* **Note 14**).
3. Centrifuge the 15 mL tube containing the resin and the bacterial soluble protein extract in a swinging bucket at 5400× *g* for 6 min at 4 °C and discard the supernatant by vacuum aspiration.
4. Fill to 13 mL with the appropriate washing buffer. Carefully resuspend the resin (*see* **Note 24**) and after centrifugation at 5400× *g* for 6 min at 4 °C, remove the supernatant by vacuum aspiration. Repeat this step twice.
5. To elute the recombinant protein, add 2.5 mL of the appropriate elution buffer to the resin, mix by gentle shaking (*see* **Note 24**) and centrifuge in a swinging bucket at 5400× *g* for 6 min at 4 °C. Collect the supernatant in a clean 15 mL tube and keep it on ice (*see* **Note 12**).

3.4 Recombinant Protein Desalting

3.4.1 Recombinant Protein Desalting by Gravity for Recombinant Protein Purified by Gravity Flow Affinity Chromatography (His_6-At*GOX1/2* and His_6-At*HPR1*)

1. Fix a PD-10 desalting column to a clamp stand and equilibrate the column (*see* **Note 12**) by gradually adding 25 mL of the appropriate desalting buffer, allowing it flow through the column and into a plastic beaker.
2. Carefully add the 2.5 mL of eluted recombinant protein from Subheading 3.3.1, **step 6** to the top of the column and let it enter by gravity flow.

3. Add 500 μL of the appropriate desalting buffer and let it flow through the column. The flow-through liquid is not kept as it does not contain recombinant protein.
4. Eluate the recombinant protein by carefully adding 3 mL of the appropriate desalting buffer and collect the eluate in a 15 mL tube kept on ice (*see* **Note 26**).

3.4.2 Recombinant Protein Desalting by Centrifugation for "Batch" Affinity Purification of Recombinant Proteins (His$_6$-AtGOX1 and His$_{10}$-AtPGLP1 and His$_6$-MBP-AtPGLP1)

1. Remove the frosted glass top filter of the PD-10 desalting column using forceps, cut the sealed end, and place it into a 50 mL collection conical tube (*see* **Note 12**) using the adapter.
2. Equilibrate the PD-10 desalting column by filling to the top with the appropriate desalting buffer until 25 mL of desalting buffer has passed through the column. Discard the flow-through liquid.
3. Centrifuge the PD-10 desalting column (placed in its collection tube with the adapter) at 1000× *g* for 2 min at 4 °C.
4. Place the equilibrated centrifuged column into a new 50 mL collection conical tube and slowly add the 2.5 mL of eluted recombinant protein from Subheading 3.3.2, **step 6** onto the equilibrated PD-10 desalting column.
5. Elute the desalted recombinant protein into the 50 mL tube by centrifugation at 1000× *g* for 2 min at 4 °C (*see* **Note 26**).

3.5 Recombinant Protein Quantification

1. Determine recombinant protein concentration by the Bradford method [20] using a commercial Bradford reagent (*see* **Note 17**) and by following the manufacturer's instructions. This will require a calibration curve made with BSA and a spectrophotometer set to 595 nm.
2. The purity of the recombinant protein can be checked by migrating 2 μg of desalted purified recombinant protein on an SDS-PAGE gel (containing the appropriate amount of acrylamide for the molecular weight of the recombinant protein) followed by staining with Coomassie blue [21] (*see* Fig. 1).

4 Notes

1. In order to have ultrapure water, we use O-TEC bi-distilled apyrogenic sterile water. The day before an experiment refrigerate some ultrapure water for the preparation of solutions and buffers. Bidistilled water was obtained using a MilliQ apparatus.
2. We use commercial LB broth (20 g/L) prepared in 1 L or 2 L beakers (depending on the required volume) with bidistilled water and sterilized in glass bottles at 121 °C for 20 min in an autoclave. Rudic culture medium (RM) is prepared as follows:

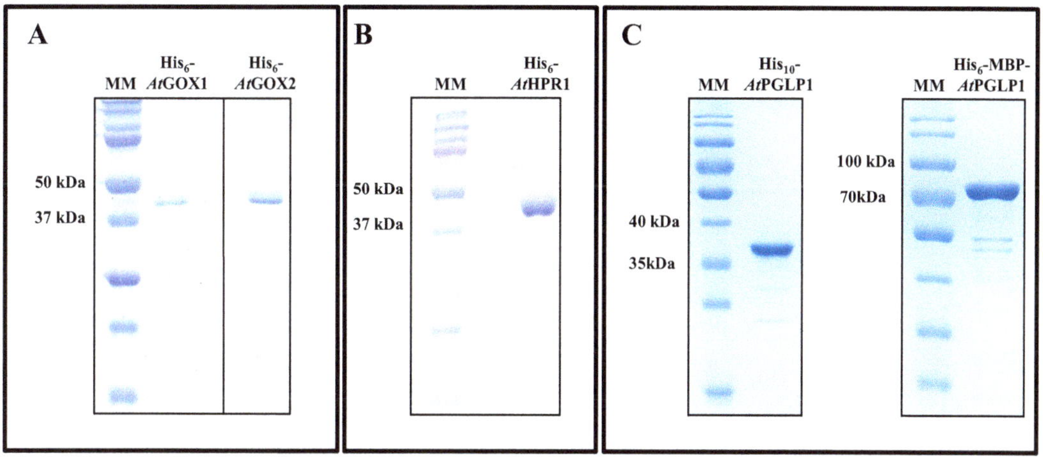

Fig. 1 Coomassie-stained purified recombinant *At*GOX1/2, *At*HPR1, and *At*PGLP1 His-tagged proteins separated on SDS-PAGE gels. (**a**) Two micrograms of desalted purified His_6-*At*GOX1 and His_6-*At*GOX2 proteins purified by gravity flow affinity chromatography were separated by SDS-PAGE with 10% acrylamide and stained with Coomassie Blue R-250. Each *At*GOX recombinant protein was estimated to be ~40 kDa. (**b**) Two micrograms of desalted purified His_6-*At*HPR1 protein purified by gravity flow affinity chromatography were separated by SDS-PAGE with 10% acrylamide and stained with Coomassie Blue R-250. *At*HPR1 recombinant protein was estimated to be ~40 kDa. (**c**) Two micrograms of desalted purified His_{10}-*At*PGLP1 and His_6-MBP-*At*PGLP1 proteins purified by "in batch" affinity chromatography were separated by SDS-PAGE either with 12% or 10% acrylamide and stained with Coomassie Blue R-250. His_{10}-*At*PGLP1 recombinant protein was estimated to be ~37 kDa and His_6-MBP-AtPGLP1 recombinant protein was estimated to be ~80 kDa. Molecular mass markers (MM) and corresponding molecular masses in kDa are shown

1× M9 salts, 2% (w/v) casamino acids, 0.2% glucose, 1 mM $MgCl_2$ in bidistilled water and sterilized at 115 °C for 20 min in an autoclave.

3. Kanamycin stock solution (50 g/L) and ampicillin stock solution (100 g/L) are prepared in bidistilled water, sterilized by filtration through a 0.25 μm filter with the help of a 50 mL syringe, aliquoted, and kept frozen at −20 °C until use.
4. 1 M IPTG stock solution is prepared in bidistilled water, sterilized by filtration through a 0.25 μm filter with the help of a 50 mL syringe, aliquoted, and kept frozen at −20 °C until use.
5. 20% (w/v) L-arabinose is prepared in bidistilled water and sterilized by filtration through a 0.25 μm filter using a 50 mL syringe, aliquoted, and kept frozen at 4 °C until use.
6. To prepare the 0.5 M Tris–HCl stock solutions at pH 7.5 and pH 8: Weigh 30.29 g of Tris-base and dissolve in a beaker in 450 mL of ultrapure water, adjust the pH to 7.5 or 8 using concentrated HCl (12.1 N) but a lower concentrated HCl solution (1 N) can be used when approaching the required pH. Then adjust the volume to 500 mL in a 500 mL graduated

cylinder with ultrapure water and homogenize and store in 250 mL glass bottles for several months at 4 °C.

7. To prepare the 3 M NaCl stock solution: Weigh 87.7 g of NaCl, dissolve it in a beaker using a magnetic stirrer in 450 mL of ultrapure water and adjust the volume to 500 mL in a 500 mL graduated cylinder with ultrapure water, homogenize and store in 250 mL glass bottles for several months at 4 °C.

8. To prepare the 50% (v/v) glycerol stock solution: Introduce directly in a graduated glass cylinder 100 mL of high purity glycerol 100% anhydrous and adjust to 200 mL with ultrapure water. Carefully seal the graduated cylinder with Parafilm M and handshake until complete homogenization. Store in a glass bottle at 4 °C for a maximum of 1 month.

9. To prepare the 0.5 M $MgCl_2$ stock solution: Weigh 20.33 g of $MgCl_2.6H_2O$ (≥99.0% purity) and dissolve it in a glass beaker with 180 mL of ultrapure water before adjusting the volume with ultrapure water to 200 mL in a 250 mL graduated cylinder, homogenize and store in a 250 mL glass bottle for several months at 4 °C.

10. 1 M imidazole stock solution is prepared freshly and not re-used therefore you should calculate the required volume before starting. Typically, to prepare 10 mL of 1 M imidazole solution: Weigh 680.6 mg of imidazole in a 15 mL conical tube and dissolve with cold ultrapure water by vortexing. Keep on ice until use.

11. To prepare 0.5 M stock solutions of Na_2HPO_4 and NaH_2PO_4: Weigh 35.5 g Na_2HPO_4 (sodium phosphate dibasic) and 34.5 g NaH_2PO_4 (sodium phosphate monobasic monohydrate) and dissolve them separately in beakers with 480 mL ultrapure water, adjust to 500 mL in a graduated cylinder, homogenize, and store in a glass bottle at 4 °C for 1 month. To prepare 100 mL of 50 mM sodium phosphate buffer pH 8 for *At*HPR1 purification: In a 100 mL graduated cylinder, add 9.32 mL of 0.5 M Na_2HPO_4, 0.68 mL of 0.5 M NaH_2PO_4, and adjust to 100 mL with ultrapure water. When required, for instance, in the case of extraction, washing, and elution buffers, add required volumes of the other constituents before adjusting to 100 mL.

12. FMN (Riboflavin 5′-monophosphate sodium salt hydrate, store at −20 °C) is unstable and sensitive to light. Thus, FMN-containing buffers must be protected from light by wrapping the bottle/tube in aluminum foil. Typically, to prepare 10 mL of 10 mM FMN stock solution: Weigh 51.4 mg of

powder in a 15 mL conical tube and resuspend in 10 mL of ultrapure water by vigourous vortexing. FMN is added to buffers just before the experiment and the stock solution is not kept.

13. We use complete EDTA-free protease inhibitor from Roche (reference: 04693159001). One tablet must be dissolved in 50 mL of buffer just before use. To be economical, a half tablet can be dissolved in 25 mL of buffer.
14. The rolling wheel is set to 8 rpm for batch and to 20 rpm for gravity-flow affinity chromatography.
15. We use HIS-Select® Nickel Affinity Gel from Sigma-Aldrich (reference: P6611).
16. We use disposable PD 10 Desalting Columns either from Cytiva (reference: 17-0851-01) or GE Healthcare (reference: GE17-0851-01).
17. For each new bottle of reagent, a new calibration curve is performed using BSA. The baseline measurement is performed in the appropriate desalting buffer without added protein.
18. Conical flask and LB medium are pre-warmed to 37 °C overnight to ensure fast bacterial growth.
19. An OD_{600nm} of 0.8 must be reached before adding the IPTG. This is important to ensure a sufficient *At*PGLP1 recombinant protein production.
20. For *At*PGLP1 and *At*GOX1/2 recombinant protein production, the bacterial pellet can be stored at −20 °C for a maximum of 3 weeks.
21. The extraction buffer must be pipetted and projected onto the pellet to dissolve it (avoid vortexing). Alternatively, a 1 mL precision pipette can be used.
22. For the first two passages through the pressure cell, the extract can be collected in the same 50 mL tube.
23. Resuspend the resin by gentle shaking. Take the appropriate volume with a 5 mL pipette and the appropriate tip or a 1 mL pipette with a cut tip to allow resin entry, otherwise only the buffer will be taken up but not the resin.
24. Resuspend the pellet of resin by gentle inversion shaking. Do not vortex.
25. The supernatant contains soluble proteins, but sometimes there is floating debris. Avoid taking this debris.
26. PD-10 desalting columns can be re-used several times, but they must be thoroughly washed with at least 25 mL of ultrapure water and stored at 4 °C with a top cap and stopper to ensure the column does not become dry.

Acknowledgments

We would like to thank Herman Bauwe and Stefan Timm (University of Rostock) for kindly providing the pBAD-HPR1 plasmid and the *E. coli* strain LMG194. We are indebted to Younès Dellero, Yanpei Liu, and Pauline Duminil who helped develop the purification protocols described in this chapter during their PhD projects carried out in the Metaboactions team at IPS2. We would like to thank the French Agence Nationale de la Recherche (ANR) for financial support: ANR-10-LABX-0040-SPS, ANR-11-IDEX-0003-02, ANR-14-CE19-0015 REGUL3P. PD and YD received PhD fellowships from the French Education Ministry and YL received a PhD fellowship from the China Scholarship Council.

References

1. Christeller JT, Tolbert NE (1978) Phosphoglycolate phosphatase. Purification and properties. J Biol Chem 253(6):1780–1785
2. Seal SN, Rose ZB (1987) Characterization of a phosphoenzyme intermediate in the reaction of phosphoglycolate phosphatase. J Biol Chem 262(28):13496–13500
3. Lindqvist Y, Brändén CI (1979) Preliminary crystallographic data for glycolate oxidase from spinach. J Biol Chem 254(15):7403–7404
4. Boldt R, Edner C, Kolukisaoglu U et al (2005) D-GLYCERATE 3-KINASE, the last unknown enzyme in the photorespiratory cycle in Arabidopsis, belongs to a novel kinase family. Plant Cell 17(8):2413–2420
5. Timm S, Nunes-Nesi A, Pärnik T et al (2008) A cytosolic pathway for the conversion of hydroxypyruvate to glycerate during photorespiration in Arabidopsis. Plant Cell 20(10):2848–2859
6. Hardy P, Baldy P (1986) Corn phosphoglycolate phosphatase: purification and properties. Planta 168(2):245–252
7. Titus DE, Hondred D, Becker WM (1983) Purification and characterization of hydroxypyruvate reductase from cucumber cotyledons. Plant Physiol 72(2):402–408
8. Studier FW, Moffatt BA (1986) Use of bacteriophage T7 RNA polymerase to direct selective high-level expression of cloned genes. J Mol Biol 189(1):113–130
9. Nallamsetty S, Waugh DS (2007) A generic protocol for the expression and purification of recombinant proteins in *Escherichia coli* using a combinatorial His6-maltose binding protein fusion tag. Nat Protoc 2(2):383–391
10. pBAD Expression System (2002) Invitrogen Life Technologies. https://assets.thermofisher.com/TFS-Assets%2FLSG%2Fbrochures%2F710_01619_pBAD_bro.pdf. Accessed 26 May 2023
11. Bornhorst JA, Falke JJ (2000) Purification of proteins using polyhistidine affinity tags. Methods Enzymol 326:245–254
12. Liu Y, Guérard F, Hodges M, Jossier M (2020) Phosphomimetic T335D mutation of hydroxypyruvate reductase 1 modifies cofactor specificity and impacts Arabidopsis growth in air. Plant Physiol 183(1):194–205
13. Timm S, Florian A, Jahnke K et al (2011) The hydroxypyruvate-reducing system in Arabidopsis: multiple enzymes for the same end. Plant Physiol 155(2):694–705
14. Schwarte S, Bauwe H (2007) Identification of the photorespiratory 2-phosphoglycolate phosphatase, PGLP1, in Arabidopsis. Plant Physiol 144(3):1580–1586
15. Hasse D, Mikkat S, Thrun HA et al (2007) Properties of recombinant glycine decarboxylase P- and H-protein subunits from the cyanobacterium Synechocystis sp. strain PCC 6803. FEBS Lett 581(7):1297–1301
16. Dellero Y, Mauve C, Boex-Fontvieille E et al (2015) Experimental evidence for a hydride transfer mechanism in plant glycolate oxidase catalysis. J Biol Chem 290(3):1689–1698
17. Jossier M, Liu Y, Massot S, Hodges M (2019) Enzymatic properties of recombinant phospho-mimetic photorespiratory glycolate oxidases from *Arabidopsis thaliana* and *Zea mays*. Plants (Basel) 9(1):27. https://doi.org/10.3390/plants9010027
18. Duminil P (2019) Characterization of two primary metabolism enzymes in *Arabidopsis*

thaliana: phosphoglycerate mutase and phosphoglycolate phosphatase. PhD dissertation, University Paris-Saclay

19. Liepman AH, Olsen LJ (2001) Peroxisomal alanine:glyoxylate aminotransferase (AGT1) is a photorespiratory enzyme with multiple substrates in *Arabidopsis thaliana*. Plant J 25(5): 487–498

20. Bradford MM (1976) A rapid and sensitive method for the quantitation of microgram quantities of protein utilizing the principle of protein-dye binding. Anal Biochem 72:248–254

21. Laemmli UK (1970) Cleavage of structural proteins during the assembly of the head of bacteriophage T4. Nature 227:680–685

Part II

Physiological Tools

Chapter 9

Measuring and Quantifying Characteristics of the Post-illumination Burst

Luke M. Gregory, Mauricio Tejera-Nieves, and Berkley J. Walker

Abstract

Leaf-level gas exchange enables accurate measurements of net CO_2 assimilation in the light, as well as CO_2 respiration in the dark. Net positive CO_2 assimilation in the light indicates that the gain of carbon by photosynthesis offsets the photorespiratory loss of CO_2 and respiration of CO_2 in the light (R_L), while the CO_2 respired in the dark is mainly attributed to respiration in the dark (R_D). Measuring the CO_2 release specifically from photorespiration in the light is challenging since net CO_2 assimilation involves three concurrent processes (the velocity of rubisco carboxylation; v_c, velocity of rubisco oxygenation; v_o, and R_L). However, by employing a rapid light-dark transient, it is possible to transiently measure some of the CO_2 release from photorespiration without the background of v_c-based assimilation in the dark. This method is commonly known as the post-illumination CO_2 burst (PIB) and results in a "burst" of CO_2 immediately after the transition to the dark. This burst can be quantitatively characterized using several approaches. Here, we describe how to set up a PIB measurement and provide some guidelines on how to analyze and interpret the data obtained using a PIB analysis application developed in *R*.

Key words C_3 photosynthesis, Net CO_2 assimilation rate, Post-illumination burst, PIB, Photorespiration, Light enhanced dark respiration

1 Introduction

Since leaf-level gas exchange only measures net CO_2 assimilation (A, full abbreviations listed in Table 1), specialized approaches are needed to estimate the component gross fluxes. The component gross fluxes associated with A in C_3 plants are CO_2 assimilation (velocity of rubisco carboxylation; v_c), the photorespiratory loss of CO_2 (usually depicted as $0.5v_o$, where v_o is equal to velocity of rubisco oxygenation), and respiration of CO_2 in the light (R_L) [1, 2]. Accurately measuring the gross CO_2 release from photorespiration in the background of CO_2 uptake can be challenging; however, it is possible to resolve values proportional to photorespiratory CO_2 release using a post-illumination CO_2 Burst (PIB) [3, 4].

Berkley J. Walker (ed.), *Photorespiration: Methods and Protocols*, Methods in Molecular Biology, vol. 2792,
https://doi.org/10.1007/978-1-0716-3802-6_9,

Table 1
List of parameters, descriptions, and their units

Parameter	Description	Unit
A	Net CO_2 assimilation rate	μmol/m^2/s
A_s	Steady state net CO_2 assimilation rate	μmol/m^2/s
g_{sw}	Stomatal conductance to water vapor	mol/m^2/s ^{1}
C_a	The CO_2 partial pressure in the ambient air	Pa
C_i	The CO_2 partial pressure in the intercellular airspace of the leaf	Pa
R_L	Non-photorespiratory release of CO_2 in the light	μmol/m^2/s
R_D	Non-photorespiratory release of CO_2 in the dark	μmol/m^2/s
T_{leaf}	Leaf temperature	°C
VPD	Vapor pressure deficit	kPa
$Burst_{Area}$	Total area of the CO_2 burst	μmol m^2
$Burst_{Time}$	Time of the CO_2 burst	minutes
Max_{Lost}	Maximum rate of CO_2 loss in the burst	μmol/m^2/s
v_c	Velocity of rubisco carboxylation	μmol/m^2/s
v_o	Velocity of rubisco oxygenation	μmol/m^2/s

The PIB method uses leaf-level gas exchange to measure A across a light-to-dark transient. When a leaf is illuminated, net CO_2 assimilation reaches a steady state with $A = v_c + 0.5v_o + R_L$, while in the dark, CO_2 respiration reaches a steady state ($A = R_D$). Consequently, a "burst" of CO_2 commonly occurs during the first few minutes immediately following the light-dark transition presumably since the reactions of the C_3 cycle cease in the darkness much faster than the CO_2 release from the photorespiratory pathway ($0.5v_o$ in the previous equation) [5, 6].

Many photorespiratory-related metabolites have been hypothesized to contribute to the PIB. Glycine is suggested to be the primary contributor of CO_2 to the burst [3, 4, 7, 8]. Glycine, a photorespiratory intermediate that accumulates in the light, continues its metabolism in the dark longer than CO_2 is assimilated, so the CO_2 released is observed in the burst. In addition to glycine, there is evidence that CO_2 release from nonenzymatic decarboxylation between H_2O_2 and hydroxypyruvate and/or glycolate in the photorespiratory pathway is also captured in the burst [9].

Other nonphotorespiratory metabolites that accumulate during photosynthesis, such as malate and pyruvate, have also been proposed to contribute to the burst. Since mitochondrial respiration is sensitive to light-dark transients, the decarboxylation of malate and/or pyruvate can stimulate initial rates of respiration in the dark after a period of illumination and has been denoted in the past as light-enhanced dark respiration (LEDR) [4, 10, 11]. Substrate feeding experiments proved evidence that glycine (200 mM) addition increased the burst size by 60% compared to water addition, while malate (200 mM) and pyruvate (200 mM) were less effective at increasing the burst size [8]. While other metabolites have been suggested, it is likely that photorespiratory intermediates play a major role in contributing CO_2 to the PIB.

Although the PIB cannot directly measure the rate of photorespiration, since there are other non-photorespiratory contributors to the PIB and the release depends on pool sizes of photorespiratory intermediates that may or may not decarboxylate at the same rate in the dark as in the light, it is still an effective method for measuring an integrated signal of carbon emptying out of the residual carbon pools in the dark. Since there is a photorespiratory component contributing to the burst, changes in the total burst size are roughly proportional to the amount of photorespiration occurring.

Along with estimates of the absolute burst size, there are additional characteristics of leaf photosynthetic and photorespiratory performance that can be estimated from PIB measurements such as:

Steady-State Net CO_2 *Assimilation Rate* (A_s): A_s is the steady state of CO_2 assimilation in the light before the leaf enters the dark and is given in units of μmol/m^2/s.

Steady-State Dark Respiration Rate (R_D): R_D is the steady-state CO_2 dark respiration rate and is given in units of μmol/m^2/s.

Area of the CO_2 *Burst* ($Burst_{Area}$): $Burst_{Area}$ is the entire amount of CO_2 released in the burst and is given in units of μmol/m^2.

Time of the photorespiratory CO_2 *burst* ($Burst_{time}$): $Burst_{time}$ is the length of time during the $Burst_{Area}$ and is given in minute units.

Maximum rate of CO_2 *loss in the burst* (Max_{lost}): Max_{lost} is the greatest rate of CO_2 lost during the $Burst_{Area}$ and is given in μmol/m^2/s.

Along with these five unique parameters, it should also be noted that there are differences between species in the PIB response. Some species (i.e., *Arabidopsis thaliana*) give a single burst, whereas other species (i.e., *Nicotiana tabacum*) give two bursts (Fig. 1). Sometimes portions of these bursts are referred to as LEDR. It is outside the scope of this method to explore the cause of these differences, but the protocol outlined below can capture and quantify these different burst types.

Here, we describe how to set up a PIB measurement, what to pay attention to while making measurements, and provide some guidelines on how to analyze and interpret the results using some quantitative methods.

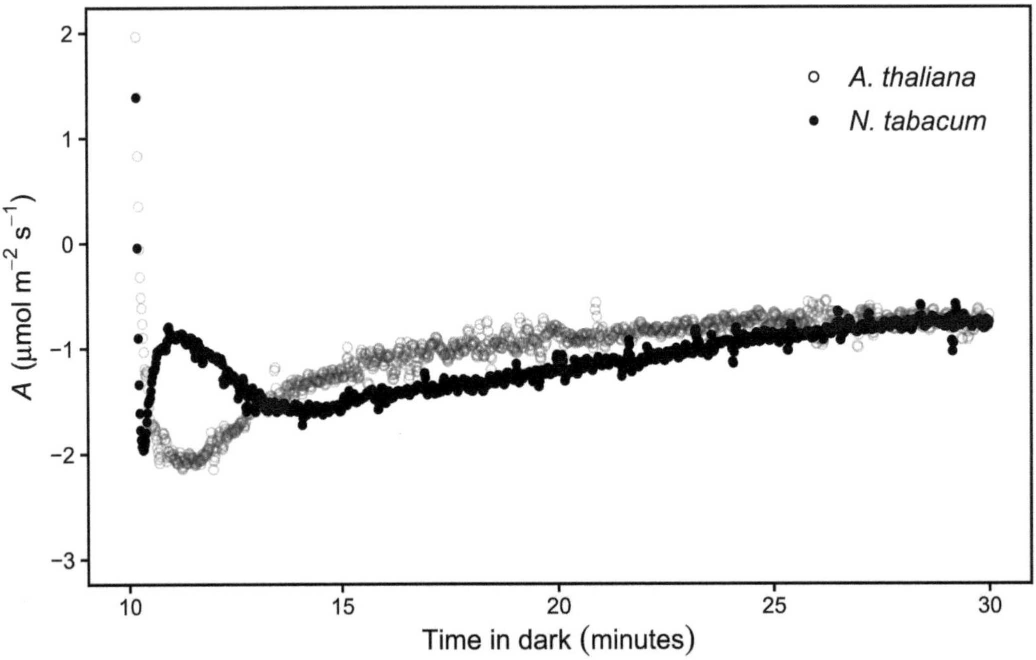

Fig. 1 Example of a single and double bursts in *Arabidopsis thaliana* and *Nicotiana tabacum*. There are two PIB behaviors that are seen across higher plant species. The first is a single burst (open points) whereby a CO_2 burst will appear immediately after the light-dark transition and will decelerate within minutes in the dark. The second PIB behavior is a double burst (closed points) whereby an initial burst of CO_2 will occur within the first minute, followed by a much longer CO_2 burst

2 Materials

1. Infrared gas exchange analyzer (IRGA) for leaves with power supply or batteries (*see* **Note 1**).
2. Light Source (*see* **Note 2**).
3. Tripod/Stand to mount the IRGA head.
4. CO_2 cartridges (*see* **Note 3**).
5. Soda lime.
6. Desiccant (such as Sorbead® Orange CHAMELEON silica gel beads or Dririte®).
7. Humidifier (If applicable for the IRGA; *see* **Note 4**).
8. Computer for data download and processing.
9. Software: ImageJ, Excel R, RStudio (Optional; *see* **Notes 5** and **6**).

3 Methods

3.1 Machine Setup and Calibration

We use the LI-6800 portable photosynthetic system for gas-exchange measurements so there may be some differences in setup based on the specific manufacturer used.

1. Before turning the IRGA system on, be sure to refill/replace the CO_2 cartridge. CO_2 scrub, Desiccant, and Humidifier chemicals of needed according to the manufacturer's instructions.
2. If needed, switch the IRGA sensor head to the appropriate leaf chamber (e.g., for the LI-6800 there is the Multiphase Flash Fluorometer, or a Large/Small Light Source that can be used). Additionally, verify that the gaskets of the leaf chamber are fitted to the IRGA sensor head properly and are not damaged. Replace damaged or ill-fitting gaskets.
3. Place the IRGA system on a clear work surface, or in a climate-controlled cabinet set to the desired measuring temperature if measuring under conditions ~ ±6 °C from ambient.
4. Power on the machine and check that the IRGA sensor head is communicating with the IRGA console. Check the temperature control on the IRGA system and set the IRGA system's desired temperature. For optimal humidity control, ensure the IRGA temperature closely matches the ambient temperature of the instrument. The IRGA itself will have a limited range of temperatures that can be achieved relative to the ambient temperature. Let the machine equilibrate for at least 45–60 min with the chamber closed. This is especially important if you are measuring above or below room temperature.
5. Perform the warm-up test (and/or the start-up checks recommended by the manufacturer). The warm-up test should reveal any warnings or errors that need attention. If there are any warnings or error messages, resolve them before beginning the experiment for the day.
6. Set the desired environmental conditions for the leaf chamber (i.e., flow rate, CO_2 and O_2 concentration, vapor pressure deficit, light intensity, and temperature).

3.2 Preparing the Leaf

1. Place the leaf into the leaf chamber on the IRGA sensor head and close the chamber fully, taking care not to fold or damage the leaf. If the leaf does not cover the entire area of the leaf gasket, you will need to measure the area of the leaf after the measurement is taken (refer to *Collect Leaf Area section* below if applicable). Ensure that the thermocouple is making firm (but not too firm as to damage the leaf) contact with the bottom of the leaf. Leaf temperature is crucial to obtain accurate

transpiration and stomatal conductance measurements. Ensure that the gaskets make a tight seal around the leaf and that there are minimal leaks.

2. Once you are confident with a secure fit onto the leaf, monitor the IRGA console. Verify that the environmental parameters are set correctly and watch the leaf parameters (i.e., A and g_{sw}) until steady-state photosynthesis is reached. Visually, steady-state photosynthesis will be reached once A and g_{sw} reach a plateau for more than 2–3 min.

3.3 Making a PIB Measurement

PIB is measured during a light-to-dark transient that can be changed manually or written in an automated program. We recommend measuring A rate for 10 min in the light and 20 min in the dark to ensure good steady states in the light and dark.

1. To establish a PIB program, create a sequence to control the light intensity of the leaf chamber. Begin the sequence by setting the desired light intensity (i.e., saturating or sub-saturating) and end the sequence by setting the light intensity to zero (i.e., dark). Set the log interval to 1 s (or the fastest the IRGA system can log measurements). Set the entire program to run over 30 min (i.e., 10 min in the light, and 20 min in the dark). Ensure that there is no additional averaging of data above the log interval.
2. Open an empty log file on the IRGA system where the data can be stored during the measurement according to the manufacturer's directions.
3. Before starting the PIB measurement, it is important to switch the humidity and temperature controls to be constant during the transient, rather than depending on leaf performance (i.e., for the LI-6800 switch VPDleaf to H2O_r and Tleaf to Txchg). It is necessary to switch instrument controls to be constant because it will not result in artifactual changes in A or g_{sw} during the transient into the dark while the instrument works to compensate for the reduction in A and radiative load.
4. After setting the humidity and temperature controls to remain constant, wait for A and g_{sw} to restabilize before starting the measurement.
5. During the measurement one of two behaviors should be present:
   (a) *Single Burst Observations*: In the light, the leaf will maintain A at a steady state. Immediately following the light-dark transition, a CO_2 burst will appear that will decelerate within minutes in the dark. After 10 min in the dark, the leaf will reach or approach an R_D rate at a steady state (Fig. 1).

(b) *Double Burst Observations*: In the light, the leaf will A at a steady state. Once the lights turn off, an initial burst of CO_2 will occur within the first minute, followed by a much longer CO_2 burst. Depending on the species that is measured, it is expected to take 15–20 min (perhaps longer, adjust program length accordingly) to reach a R_D rate at a steady state (Fig. 1).

3.4 Collect Leaf Area (When Necessary)

1. Once the PIB measurement is finished, mark the leaf with a marker where the gaskets are clamped on.
2. Release the leaf from the leaf chamber. Place a spare gasket overtop of the leaf and line it up with the marked spots. Place a known scale in the image field, such as a ruler, and take a picture of the leaf with the gasket placed overtop (ideally on a white surface).
3. Using a digital area analyzer, like ImageJ, determine the leaf area inside of the gasket [12].
4. Record the leaf area and correct the gas exchange data accordingly before analyzing the data (column "S" on LICOR Excel file).

3.5 Analyzing and Interpreting the PIB Data

The previous sections were to ensure that the highest quality of data is obtained for the PIB. The following section will provide insight into how to analyze and interpret the PIB data (Fig. 2).

1. Plot A as a function of time and assess the quality of the data.
2. Perform initial quality control on the data. Visually, is there a lot of noise in the measurement? Does A rate reach a steady state in the light? Is there a distinct CO_2 burst immediately following the light-dark transition? Does R_D reach a steady state in the dark? If the data is satisfactory, then the PIB parameters can be estimated from the data.
3. To estimate A_s and R_D, average the last few points before the light transition, and the last few points in the dark before the measurement ends, respectively.
4. Add a linear regression overtop of the CO_2 burst, whereby the y-intercept is the solved $R_{D,}$ and the slope is set to zero.
5. To estimate $Burst_{Area}$, sum up the differences between the R_D linear regression and the measured assimilation values within the burst.
6. To estimate $Burst_{Time}$, find the difference between the first and last time point identified as the CO_2 burst.
7. To estimate Max_{Lost}, find the length from the R_D linear regression to the minimum assimilation in the CO_2 burst.

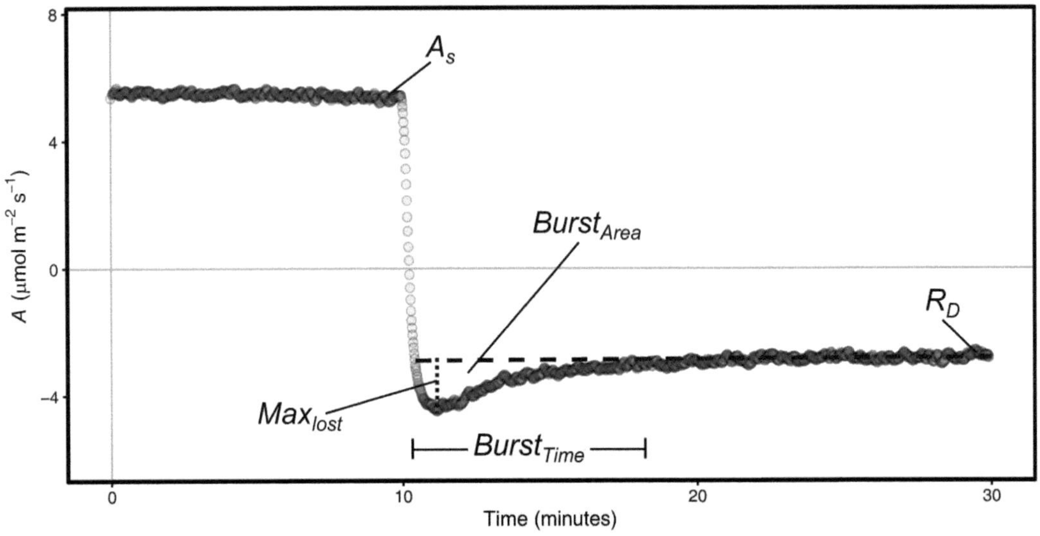

Fig. 2 Estimated parameters overlayed onto post-CO_2 illumination burst. There are five unique parameters (A_s, R_D, $Burst_{Area}$, $Burst_{Time}$, Max_{lost}) that provide characteristics of leaf photosynthetic and photorespiratory performance that can be estimated from a PIB. The location of where these parameters are estimated is labeled onto of the PIB

3.6 Using the ShinyApp to Automate Analysis

The previous sections described how to check PIB data collected for quality, and how to analyze and interpret the PIB parameters manually. The following section introduces a custom-built program (i.e., ShinyApp) that can automate the analysis used to determine the PIB parameters (Fig. 3).

1. Download the custom-built program from GitHub (https://github.com/L-Gregory/PIB_analyzer).
2. Open the script in RStudio and select Run App, this will open the application in another RStudio window.
3. Upload data by selecting the Browse button, or by "dragging and dropping" the data file into the box. Data must contain net assimilation rate and time columns.
4. Once data is uploaded successfully, select the columns containing time and assimilation. This will plot the data.
5. The data can be visually checked for quality now.
6. If the data are acceptable, click the "Fit" button. This will automatically estimate and display the five parameters of interest (A_s, R_D, $Burst_{Area}$, $Burst_{Time}$, Max_{Lost}) below the plot.
7. If you would like to change the R_D estimate for the linear regression—check the "Change Rd fit" box. Using the slider bar, select the number of points that you want to be included in the linear regression (default = 50).

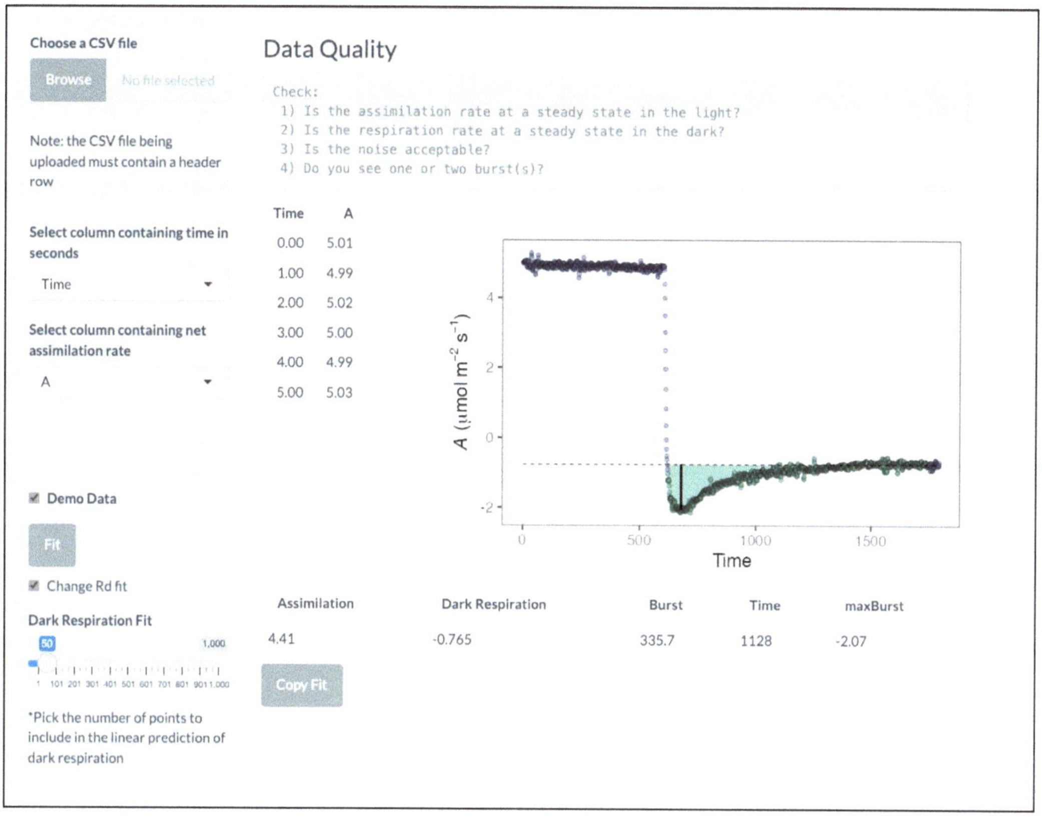

Fig. 3 Custom-written application for analyzing post-illumination CO_2 burst data. Screen-shot from the custom-built program that automates estimation of the five unique parameters (A_S, R_D, $Burst_{Area}$, $Burst_{Time}$, Max_{lost}) from the uploaded PIB data. Upon executing the app, PIB data can be uploaded into the file selection box. After defining the axes, a plot of the data will appear in the main panel. Clicking on the "Fit" button will generate the five parameters of interest. If the linear regression does not fit properly, the user can adjust the points that are included in it (default = 50)

4 Notes

1. This method described here refers to gas exchange measurements made on the LI-6800 portable photosynthetic system (LICOR, Lincoln, NE, USA). The method could be adjusted for instruments from other manufacturers that measure CO_2 and H_2O gas exchange with high temporal resolution, as the general principle of operation, data collection, and analysis/interpretation still applies. However, we have not checked this method performance on other systems.

2. The simplest option is the LED light source made specifically for the leaf chamber in use, designed by the IRGA manufacturer. Depending on experimental requirements other light sources may be required (e.g., when a custom-built or clear

top chamber is used, or when a certain light quality is required).

3. CO_2 cartridges should be purchased from the manufacturer if there is not a replaceable prefilter. Use of generically branded CO_2 cartridges can introduce oil into the system and damage the instrument.
4. If you have a Nafion tube humidifier column, use deionized H_2O sources, instead of non-deionized H_2O as this can cause mineral buildup over time in the instrument.
5. Digital analysis for leaf area can be accomplished using ImageJ. However, this could be adapted to another application quite easily.
6. R code is provided in the supplemental section that automatically analyzes and estimates the PIB parameters given PIB dataset.

References

1. Farquhar GD, von Caemmerer S, Berry JA (1980) A biochemical model of photosynthetic CO2 assimilation in leaves of C3 species. Planta 149(1):78–90
2. von Caemmerer S (2013) Steady-state models of photosynthesis. Plant Cell Environ 36(9): 1617–1630
3. Busch FA (2013) Current methods for estimating the rate of photorespiration in leaves. Plant Biol 15(4):648–655
4. Sharkey TD (1988) Estimating the rate of photorespiration in leaves. Physiol Plant 73(1): 147–152
5. Decker JP (1955) A rapid, postillumination deceleration of respiration in green leaves. Plant Physiol 30(1):82–84
6. Decker JP (1959) Comparative responses of carbon dioxide outburst and uptake in tobacco. Plant Physiol 34(2):100–102
7. Rawsthorne S, Hylton CM (1991) The relationship between the post-illumination CO_2 burst and glycine metabolism in leaves of C3 and C3-C4 intermediate species of Moricandia. Planta 186(1):122–126
8. Parys E, Romanowska E (2000) Relationship between postillumination burst of CO_2 and enhancement of respiration in tall fescue leaves. Acta Physiol Plant 22(2):135–142
9. Bao H et al (2021) Catalase protects against nonenzymatic decarboxylations during photorespiration in *Arabidopsis thaliana*. Plant Direct 5(12):e366
10. Azcón-Bieto J, Osmond CB (1983) Relationship between photosynthesis and respiration: the effect of carbohydrate status on the rate of CO(2) production by respiration in darkened and illuminated wheat leaves. Plant Physiol 71(3):574–581
11. Atkin OK, Evans JR, Siebke K (1998) Relationship between the inhibition of leaf respiration by light and enhancement of leaf dark respiration following light treatment. Funct Plant Biol 25(4):437–443
12. Schneider CA, Rasband WS, Eliceiri KW (2012) NIH image to ImageJ: 25 years of image analysis. Nat Methods 9(7):671–675

Chapter 10

Low-Cost System for Maintaining Low Atmospheric CO_2 Levels During Arabidopsis Cultivation

Naveen Sharma and Federica Brandizzi

Abstract

Photosynthesis requires CO_2 as the carbon source, and the levels of ambient CO_2 determine the oxygenation or carboxylation of Ribulose-1,5-bisphosphate (RuBP) by RuBP carboxylase/oxygenase (Rubisco). Low CO_2 levels lead to oxygenation and result in photorespiration, which ultimately causes a reduction in net carbon assimilation through photosynthesis. Therefore, an increased understanding of plant responses to low CO_2 contributes to the knowledge of how plants circumvent the harmful effects of photorespiration. Methods for elevating CO_2 above ambient concentrations are often achieved by external sources of CO_2, but reducing CO_2 below the ambient value is much more difficult as CO_2 gas needs to be scrubbed from the atmosphere rather than added to it. Here, we describe a low-cost method of achieving low CO_2 conditions for Arabidopsis growth.

Key words Arabidopsis, Low-CO_2, Photorespiration

1 Introduction

Photosynthesis is the process by which light energy is captured by plants and used to synthesize reduced carbon compounds from carbon dioxide and water. Ribulose-1,5-bisphosphate (RuBP) carboxylase/oxygenase (RubiscO) is the carbon-fixing enzyme present in most photosynthetic organisms, converting CO_2 into organic matter. However, the primary substrate (CO_2) competes with O_2 for binding to the RuBP in the Rubisco active site [1]. Low CO_2 causes reduced photosynthesis and increased photorespiration, which leads to reduced production of carbohydrates in plants [2]. To circumvent the negative effects of photorespiration, many photosynthetic organisms have evolved and acquired metabolic and spatial mechanisms either to maximize CO_2 concentration around Rubisco or to minimize the loss of photorespired CO_2. Therefore, an increased understanding of how photosynthetic organisms respond to low CO_2 contributes to the knowledge of how plants

Berkley J. Walker (ed.), *Photorespiration: Methods and Protocols*, Methods in Molecular Biology, vol. 2792,
https://doi.org/10.1007/978-1-0716-3802-6_10,

circumvent the negative effects of photorespiration. Low CO_2 conditions can also be used to identify mutants in the photorespiratory pathway and screening for the photorespiratory bypasses [3, 4]. Moreover, such understanding will contribute to the knowledge of how natural global change factors in the past may continue to influence plant responses to future anthropogenic changes [5]. While methods for elevating CO_2 above ambient concentrations are often achieved via external sources of CO_2 and can easily be applied to growth chambers, reducing CO_2 levels below the ambient level is much more difficult and expensive [5]. Thus, we have devised a low-cost system of maintaining low CO_2 for Arabidopsis growth by employing commonly used materials. Low CO_2 treatments are imposed most commonly for entire chambers by scrubbing the ambient CO_2, either by passing the air through "scrub boxes" containing soda lime or by-passing compressed air across soda lime filters. These methods are effective for entire growth chambers but are expensive and not widely available. Here, we used a standard growth tray and added soda lime for scrubbing of CO_2 within the growth tray. The growth tray is coupled with humidity vents, and adjusting the vent size allows reaching the desired CO_2 levels.

2 Material

1. Growth tray with humidity vents ($15 \times 9.4 \times 7.09$ inches).
2. Soda lime.
3. Multipurpose laboratory tape.
4. 25% (v/v) bleach solution.
5. Murashige and Skoog basal salts + macro and micronutrients.
6. 0.1% sucrose.
7. Petri plates.
8. Plant agar.
9. Arabidopsis seeds.
10. Infra-Red Gas analyzer (to measure CO_2).
11. Vortex.
12. Potassium hydroxide (KOH).

3 Methods

1. Collect small aliquots of desired Arabidopsis seeds into clean microfuge tubes. Surface sterilize seeds with 1 mL of a solution of 25% (v/v) household bleach by vortexing for 10 min.

Fig. 1 Picture of setup used for low CO_2 cultivation of Arabidopsis seedlings

Table 1
CO_2 level inside growing tray on first day of treatment

S. no.	Time (min)	CO_2 concentration (ppm)	Humidity vent
1	0	10	Closed
2	10	22	Open
3	20	28	Open
4	30	88	Open
5	40	80	Open
6	50	85	Open
7	60	80	Open

Discard the supernatant and rinse with distilled water at least three times. Plate seeds on a medium containing half-strength Murashige and Skoog nutrients, pH 5.7, solidified with agar, and stratify them for 3 days (*see* **Notes 1**and **2**).

2. Weigh 100 grams of soda lime and place it in the bottom of growth tray with humidity vents (*see* **Note 3**).
3. Depending upon the experiment, place the plated seed plates or 10-day seedlings in the growth trays.
4. Seal the growth tray with multi-purpose laboratory tape and measure the CO_2 levels in the sealed chamber using an Infra-Red Gas analyzer (Fig. 1) (*see* **Note 4**).
5. Adjust the vent aperture to stabilize to the desired CO_2 levels in the tray.
6. Monitor the CO_2 levels till the CO_2 level stabilizes in the growth tray (Tables 1 and 2).

Table 2
CO_2 level inside growing tray on day second day of treatment

S. no.	Time (min)	CO_2 concentration (ppm)	Humidity vent
1	0	82	Open
2	10	81	Open
3	20	84	Open
4	30	80	Open
5	40	79	Open
6	50	80	Open
7	60	83	Open

4 Notes

1. The seeds can be grown on pots for giving low CO_2 treatment to 1-month-old Arabidopsis plants.
2. Place the control plates and low-CO_2 treatment container within the same growth chamber.
3. 100 grams of soda lime can maintain approximately 80 ppm in a growth tray (15 × 9.4 × 7.09 inches) with one of the vents open (Table 2).
4. LICOR 6800 can be used for monitoring the CO_2 levels inside the container by inserting one of the tubes connecting LICOR with the growth tray through the humidity vent.

Acknowledgments

We would like to thank Dr. Berkley J. Walker (Michigan State University) for his assistance in developing this protocol. This work was supported by the Chemical Sciences, Geosciences and Biosciences Division, Office of Basic Energy Sciences, Office of Science, US Department of Energy (award number DE-FG02-91ER20021), and AgBioResearch (MICL02598).

References

1. Cummins PL (2021) The coevolution of RuBisCO, photorespiration, and carbon concentrating mechanisms in higher plants. Front Plant Sci 12:662425
2. Peterhansel C, Maurino VG (2011) Photorespiration redesigned. Plant Physiol 155:49–55
3. Badger MR, Fallahi H, Kaines S, Takahashi S (2009) Chlorophyll fluorescence screening of Arabidopsis thaliana for CO(2) sensitive photorespiration and photoinhibition mutants. Funct Plant Biol 36:867–873
4. South PF, Cavanagh AP, Liu HW, Ort DR (2019) Synthetic glycolate metabolism pathways stimulate crop growth and productivity in the field. Science 363:eaat9077
5. Gerhart LM, Ward JK (2010) Plant responses to low [CO_2] of the past. New Phytol 188:674–695

Chapter 11

Measuring the Release of Ammonia from Leaves

Shin-Ichi Miyazawa

Abstract

Ammonia (NH_3) is released from the leaves to the atmosphere when atmospheric NH_3 concentration is low; in contrast, when atmospheric NH_3 concentration is high, NH_3 in the atmosphere is absorbed by the leaves. Some previous studies have examined relationships of such NH_3 gas exchange with photorespiration, because a NH_3 production reaction is involved in the photorespiratory pathway. NH_3 compensation point (χ_{NH3}) is known as a parameter that represents an NH_3 emission potential of the leaves. Two main procedures for determining the χ_{NH3}: "gas exchange method" and "apoplast extraction method" are explained in this chapter.

Key words Ammonia, Ammonia compensation point, Apoplast, Gas exchange

1 Introduction

NH_3 is exchanged between the leaves and the atmosphere (leaf NH_3 gas exchange). When atmospheric NH_3 concentration is low, a part of the NH_3 in leaf cells is transferred to the cell walls and diffuses out to the atmosphere through stomata; on the contrary, when atmospheric NH_3 concentration is high, NH_3 in the atmosphere is absorbed by the leaves. During photorespiration, NH_3 is produced by the reactions of the glycine decarboxylase multienzyme complex in the mitochondria (Fig. 1). The NH_3 is transferred by a transporter or simply diffuses through the cytosol to the NH_3 assimilation sites, chloroplasts. The rates of leaf photorespiratory NH_3 production probably increase when the leaves are exposed to high O_2 or low CO_2 concentrations, because the photorespiratory activities are enhanced under such conditions. In previous studies, the changes in the rates of NH_3 release from the leaves subjected to different O_2 concentrations had been examined for oilseed rape, sorghum, soybean, rice, and wheat [1–4]; those subjected to different CO_2 concentrations were reported only for barley and rice [5, 6]. These reports have suggested that the rates of NH_3 release did not simply increase with the increases in the

Berkley J. Walker (ed.), *Photorespiration: Methods and Protocols*, Methods in Molecular Biology, vol. 2792,
https://doi.org/10.1007/978-1-0716-3802-6_11,

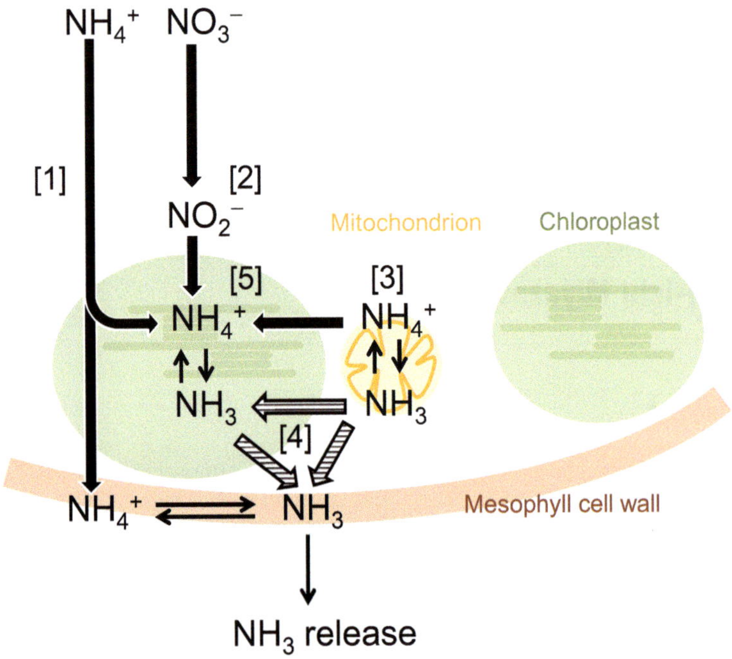

Fig. 1 Physiological factors affecting the rates of NH_3 release from illuminated mature leaves: [1] amounts of NH_4^+ transported via the xylem sap stream; [2] nitrate assimilation rates; [3] NH_3 production rates by glycine decarboxylase multienzyme complex; [4] rates of NH_3 transport to the cell walls across the organelles, cytosols and plasma membranes via transporters and/or a simple diffusion (striped arrows); [5] NH_3 assimilation rates by glutamine synthetase/ glutamate–2-oxoglutarate aminotransferase (GS/GOGAT) cycle. NH_4^+transported via the xylem sap stream is assimilated by GS/GOGAT and a part of that remains could be transported to the mesophyll cell walls. In senescing leaves, the rates of NH_3 that is liberated during protein degradation should be included [8]

rates of photorespiratory NH_3 production. Furthermore, in rice, the NH_3 release responses were different between O_2 and CO_2 treatments [4, 6]. Therefore, the responses are also likely influenced by factors such as nitrate assimilation, the location of mitochondria within the mesophyll cell, and other species-specific differences ([6]; Fig. 1).

Leaf NH_3 gas exchange rate per unit leaf area (F_{NH3}) linearly increases with the increase in NH_3 partial pressures in the air ($[NH_3]_{air}$) with an *x*-intercept (Fig. 2). This was first reported by Farquhar et al. [7], who analyzed the NH_3 gas exchange rates for French bean (*Phaseolus vulgaris*) of which aerial parts were enclosed in a glass-made chamber and were exposed to different NH_3 concentrations in air entering the chamber. The *x*-intercepts are known as the NH_3 compensation points (χ_{NH3}; nmol/mol, nbar or Pa in pressure units), an index of NH_3 emission potential of plants. The slope of the linear relationships between F_{NH3} and $[NH_3]_{air}$ is often

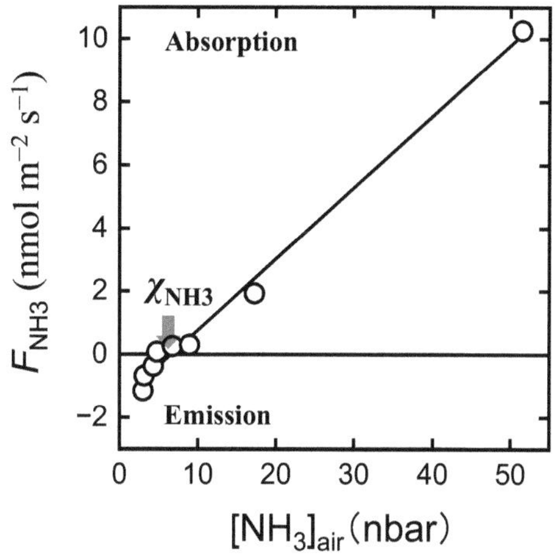

Fig. 2 Relationship between NH_3 gas exchange rates per unit leaf area (F_{NH3}) and NH_3 partial pressures in the air ($[NH_3]_{air}$) entering the chamber in *Phaseolus vulgaris* (measured at 33.4 °C). (The data are taken from [7])

assumed to be proportional to stomatal conductance to water vapor [8].

The χ_{NH3} increases with the increases in NH_4^+ concentrations of the leaf apoplast solutions ($[NH_4^+]_{apo}$, [7]). The increases in temperature and pH of the leaf apoplast solutions also increase the χ_{NH3} because gaseous NH_3 becomes undissolved from a solution under high temperature and alkaline conditions [7]. The χ_{NH3} values are determined by two approaches: the "gas exchange method" and "apoplast extraction method." The gas exchange method requires the measurements of F_{NH3} with various $[NH_3]_{air}$ surrounding the leaves using a chamber system: the χ_{NH3} is determined from the values of x-intercepts crossed with the linear relationships between F_{NH3} and $[NH_3]_{air}$. For the apoplast extraction method, the apoplast solution is extracted from the leaves with a centrifuge and the χ_{NH3} is determined from the measured NH_4^+ concentrations and pH of the extracted apoplast solutions.

2 Materials

2.1 Gas Exchange Method

Figure 3 outlines a chamber system used for NH_3 gas exchange measurements of rice (*Oryza sativa*) leaves exposed to controlled-CO_2 concentrations [6, 9]. NH_3 in the air exhausted from leaf chambers is trapped in cellulose filter paper impregnated with phosphoric acid/glycerol solution (*see* **Note 1**). The use of polytetrafluoroethylene (PTFE) tubing is recommended for constructing the system to reduce the adhesion of NH_3 on tubing. The numerical numbers listed below correspond to those depicted in Fig. 3a.

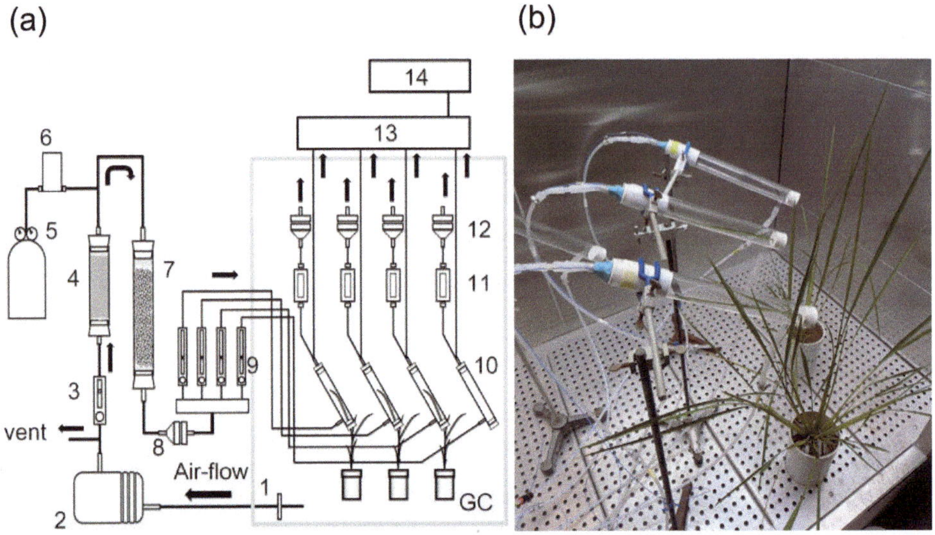

Fig. 3 Outline of a custom-made chamber system for NH_3 gas exchange measurements (**a**) and a photograph of the chamber system in a growth cabinet (**b**). (Figure 3a is reproduced with permission from "Oxford University Press" [6])

1. PTFE filter (pore size = 1 μm).
2. Air Pump with the maximum flow rate of 15 L/min.
3. Rotameter with the maximum flow rate of 30 L/min.
4. CO_2 absorption column. Acrylic column (0.4 L in volume) containing soda lime to absorb CO_2 in the air. Soda lime is sandwiched between the layers of glass wool inside the column.
5. CO_2 cylinder (99.999%).
6. Mass-flow controller.
7. NH_3 absorption column. Acrylic column (1 L in volume) is filled with 1 kg glass beads (3 mm diameter), which have been coated with 50 mL solution containing 2% (v/v) phosphoric acid and 5% (v/v) glycerol and dried for 4 days at 60 °C in an oven. The glass beads are sandwiched between the layers of glass wool inside the column.
8. NH_3 absorption filter. Cellulose filter paper (45 mm in diameter), which has been impregnated with the above phosphoric acid/glycerol solution and dried, is set into a filter holder (Two-stage inline type). A sheet of the NH_3 absorption filter papers is mounted on each stage of the filter holder (two sheets in total).
9. Rotameters with the maximum flow rate of 5 L/min.
10. Acrylic chambers with silicon foam plugs at both ends (*see* **Note 2**). Transparent cylindrical chamber (40 × 3.6 cm in

length and diameter) is used for rice leaf blades. The plug at the bottom end is slit to allow insertion of the laminas (*see* **Note 3**).

11. Mass-flow meters.
12. NH_3 trapping filters. Cellulose filter paper (30 mm in diameter) that has been impregnated with the above phosphoric acid/glycerol solution and dried is set in a filter holder (two-stage inline type) [10]. A sheet of PTFE filter is installed at the upstream stage of the holder for filtering out aerosols; a sheet of the NH_3 trapping filter paper is installed at the downstream stage.
13. Switching manifold that allows analysis of the gas from one lamina at a time.
14. Infrared gas analyzer.

2.2 Apoplast Extraction Method

1. Sorbitol (or mannitol) solution containing 0.01% Tween 20 (v/v). The concentrations of sorbitol (or mannitol) solutions are determined from the osmotic pressure of sap squeezed from the leaves harvested during the daytime using an osmometer (*see* **Note 4**).
2. 20 mM formate solution.
3. Razor blades.
4. Vacuum desiccator with a pump.
5. Lint-free tissue paper.
6. Glass slides.
7. Semimicro pH meter, an accurate and adequate for a small quantity of liquid samples.
8. Tweezers.
9. Centrifugal filter device with pore size of 0.45 μm.
10. Centrifuge (centrifugation at ×2000 *g* is required).
11. 2 mL microtubes.
12. Micropipette.

2.3 NH_4^+ Determinations by *o*-Phthalaldehyde (OPA) Method

2.3.1 Reagents

1. The OPA reagent [phosphate buffer solution containing 10 mM 2-mercaptoethanol (2-ME) and 3 mM *o*-phthalaldehyde (OPA)] is prepared with:

   Solution A: Phosphate buffer stock solution:

   Prepare 0.5 M KH_2PO_4 and 0.5 M K_2HPO_4 solutions. Mix both solutions in an equal volume (100 mL each). The pH of this solution should be approximately 6.7 at room temperature.

   Solution B: OPA solution:

0.2 g OPA (suitable for fluorometry) is dissolved in 2.5 mL of ethanol. This solution B should be prepared immediately prior to use (*see* **Note 5**).

200 mL of solution A is diluted with distilled water to a volume of approximately 450 mL. Add 350 μL 2-ME and solution B in this diluted solution. The mixed solution is diluted with distilled water to a volume of 500 mL with a graduated measuring cylinder.

2. NH_4^+ standard.

   Solution C: 100 mM NH_4Cl stock solution:

   0.107 g NH_4Cl is dissolved in 20 mL of 5 mM formate solution (can be stored at 4 °C, **Note 6**). 10 μL of solution C is diluted with distilled water to a volume of 1 mL prior to use (1 mM NH_4^+ in final concentration).

2.3.2 Outline of a Column-Less HPLC System

The numerical numbers listed below correspond to those depicted in Fig. 4. An HPLC equipped with a fluorescence detector is used:

1. OPA-reagent.
2. HPLC pump.
3. HPLC injector (less than 100 μL).
4. Water bath.
5. Stainless steel tubing. A part of the stainless-steel tubing is coiled and the coiled part is immersed in the water bath for proceeding with the reactions of NH_4^+ in the samples with OPA for 3 min at 60–65 °C; another part of the tubing is coiled outside the water bath for cooling the OPA reagent before entering the fluorescence detector.
6. HPLC fluorescence detector (excitation and emission wavelengths for fluorescence measurements are set at 410 and 470 nm, respectively).

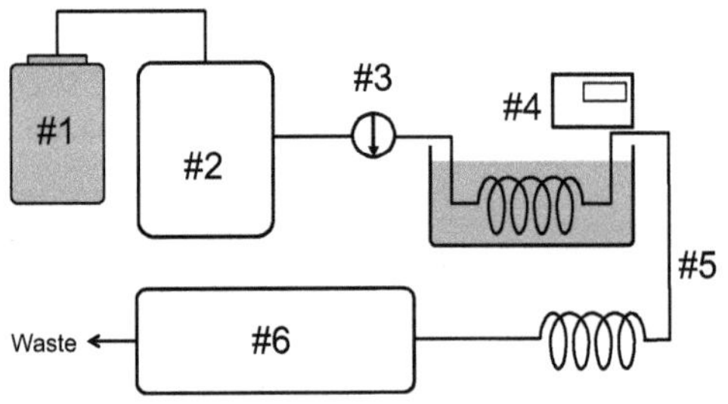

Fig. 4 Flow diagram of a column-less HPLC for NH_4^+ measurements

3 Methods

3.1 Gas Exchange Method

The numerical numbers with the hash symbol correspond to the elements depicted in Fig. 3a.

3.1.1 Measurement Procedures Using a Custom-Made Chamber

1. Turn on the pump (#2). Circulate NH_3-free air (the air passing through the NH_3 absorption column and filter) through the NH_3 gas exchange system for at least 12 h.
2. Transfer plants to the environmentally controlled growth cabinet (GC) and illuminate at a desired light intensity.
3. Install leaf blades into the chambers. At least one chamber is used as a control (i.e., the chambers without leaf blades).
4. CO_2 concentrations in the system are adjusted using the mass-flow controller (#6) and the rotameter (#3).
5. Connect the chamber exhaust tubes to the switching manifold (#13).
6. Adjust the flow rate entering each chamber to 3 L/min (#9).
7. Wait for approximately 30 min to 1 h until the photosynthesis rates are at steady states.
8. Monitor CO_2 and water vapor concentrations in the air leaving each chamber for about 5 min using the IRGA (#14).
9. Disconnect the exhaust tubes of the chambers and the switching manifold.
10. Connect the exhaust tubes to the filter holders of which newly prepared PTFE and NH_3 trapping filters are already installed (#12).
11. Check the flow rates of the rotameters (#9) adjust them entering the chambers again, and record their flow rates.
12. Record the air temperatures inside the chambers.
13. Wait for more than 60 min to trap NH_3 on the NH_3 trapping filter papers (*see* **Note** 7). During NH_3 trapping, record the mass flow rates (#11).
14. After NH_3 trapping, disconnect the filter holders and the exhaust tubes. Open the filter holder lids and collect the NH_3 trapping filter papers carefully with a pair of tweezers.
15. Immerse each NH_3 trapping filter paper into 1.0 mL of distilled water in 1.5 mL microtubes.
16. Dissolve the NH_3 trapped in the filter paper and determine the concentration using an appropriate method (e.g., with the OPA method that is described below). NH_4^+ concentrations of some blank filter papers (i.e., the filter papers only impregnated with the phosphoric acid/glycerol solutions) also should be determined.

17. Remove the NH_3 absorption column and filter (#7 and #8) to allow the ambient air to circulate through the system (*see* **Note 8**).
18. Repeat **steps 10–16**. This allows measurement of the NH_3 gas exchange rates at an increased NH_3 concentration surrounding the laminas.
19. Remove the laminas from the chambers. Measure the areas of the lamina that had been treated inside the chambers with a leaf area meter or with a scanner.

3.1.2 Calculating χ_{NH3} from NH_3 Gas Exchange Data

The rates of net CO_2 fixation per unit leaf area (P_n) is calculated according to the following equation [11]:

$$P_n = (c_o' - c_o)\tilde{n}u_e/\mathrm{LA} - c_o\tilde{n}E$$

where c_o' and c_o are the mole fractions of CO_2 in air leaving from the control (without lamina) and sample (lamina inside) chambers, respectively. u_e (mol/min), an average molar flow rate entering the chambers, is calculated from the air temperatures inside the chambers and the flow rates measured with the rotameters (#9). LA (m^2) and E are the lamina area and the transpiration rate per unit leaf area, respectively. E is determined as

$$E = (w_o - w_o')\tilde{n}u_e/((1 - w_o)\tilde{n}\mathrm{LA})$$

where w_o and w_o' are the mole fractions of water vapor in air (mol/mol) leaving the sample and control chambers, respectively. Leaf NH_3 gas exchange rate per unit leaf area (F_{NH3}) is calculated using the following equation:

$$F_{NH3} = (n_o' - n_o)\tilde{n}u_e/\mathrm{LA} - n_o\tilde{n}E$$

where n_o' and n_o are the molar fractions of NH_3 in the air leaving the control and sample chambers, respectively. n_o is calculated from the following equation:

$$n_o = N/(u_o\tilde{n}t)$$

where N, u_o, and t are the total amounts of NH_4^+ (mol) extracted from the NH_3 trapping filter papers (#12), the molar flow rates (mol/min) measured with a mass-flow meter (#11), and the sampling periods (min), respectively. The N is determined after subtraction of the NH_4^+ amounts in blank filter papers. n_o' is determined in the same way. Plotting F_{NH3} values against n_o' should provide a linear relationship with an x-intercept. The x-intercepts correspond to the χ_{NH3}.

When total leaf resistance to water vapor (r_{tot} = stomatal resistance + boundary layer resistance) and the molar fractions of NH_3 surrounding the leaves (n_a, mol/mol) are correctly determined, χ_{NH3} values also can be estimated from the following equation [8, 12]:

$$\chi_{NH3} = n_a - 1.087\tilde{n}r_{tot}\tilde{n}F_{NH3}$$

The coefficient 1.087 is the diffusivity ratio of water vapor to NH_3. The χ_{NH3} values are expressed as a partial pressure at a given atmospheric pressure.

3.2 Apoplast Extraction Method

3.2.1 Extraction Procedures for Leaf Apoplast Solutions

The apoplast solutions from the laminas are extracted by means of vacuum infiltration, which is based on the methods described by Husted and Schjoerring [13]. The method described here was developed for rice (*Oryza sativa*) leaf blades [6].

1. Cut lamina (approximately 10 cm in length) into three or four parts with a clean, sharp razor blade.
2. Immerse each part of the lamina into 2.0 mL isotonic sorbitol solution containing 0.01% (v/v) Tween 20 in a 2 mL microtube.
3. Place the microtubes (keep the microtube lids open) into a vacuum desiccator chamber connected with a pump. Draw the air from the desiccator, wait until bubbles appear on the lamina surfaces, and then return to atmospheric pressure.
4. Repeat **step 3** several times until the leaves are infiltrated with the sorbitol solutions (the laminas become dark green when those are infiltrated).
5. Immediately after the infiltration, blot the lamina surface dry with tissue paper.
6. Cut the lamina into segments (approximately 1 cm in length) with a clean, sharp razor blade, and place a bunch of their segments into a centrifugal filter device with the cut surfaces downward.
7. After centrifugation at 2000 × g for 30 s at room temperature (*see* **Note 9**), the resultant filtrates are taken as the apoplast solutions (*see* **Note 10**).
8. Drop the extracted apoplast solutions on a clean glass slide with a micropipette. Immediately measure the pH of the apoplast solutions with a semimicro pH meter at room temperature.
9. After measuring the pH, collect the apoplast solutions and return to the microtubes with a micropipette (record the amounts of the collected apoplast solutions).
10. Dilute the apoplast solutions with an appropriate amount of 20 mM formate solution (10 mM in final concentration) and keep the diluted solutions at 4 °C until NH_4^+ measurements (*see* **Note 6**).
11. Determine the NH_4^+ concentration of the apoplast solutions with the OPA method that is described below (*see* **Note 11**).

3.2.2 Calculating χ_{NH3} for the Apoplast Solutions

Calculate the χ_{NH3} from the NH_4^+ concentrations and the pH of the apoplast solutions using the equation provided by Farquhar et al. [7]:

$$\chi_{NH3} = R\tilde{n}T\tilde{n}\left(10^{(1.60352-4207.62/T)}\right)\tilde{n}\left[NH_4^+\right]_{apo}/[H^+]_{apo}$$

where R is the gas constant (= 0.0831 (L bar)/K). T, $[NH_4^+]_{apo}$, and $[H^+]_{apo}$ are the temperature (K), the apoplastic NH_4^+, and proton concentrations (mol/L), respectively (*see* **Note 12**).

3.3 Determining NH_4^+ by the o-Phthalaldehyde (OPA) Method

The OPA method is based on the method using a column-less HPLC system originally developed by Goyal et al. [14]. The OPA method is highly sensitive and selective compared with the other methods for NH_4^+ determinations [15]. The numerical numbers with the hash symbol correspond to the elements depicted in Fig. 4.

1. Set the flow rate of the HPLC pump (#2) and the temperature of the water bath (#4) at 1 mL/min and 63 °C, respectively.
2. Wait until the fluorescence signal becomes stable.
3. Inject 20 μL of standards or samples (#3). The fluorescence peaks of the NH_4^+–OPA products should be detected.
4. Calculate the fluorescence peak areas.
5. Obtain a linear calibration line between the areas and the NH_4^+ concentrations of the standard series using a diluted series of 1 mM NH_4^+ standard.
6. NH_4^+ concentration of the samples is determined from the sample peak areas using the calibration standard curve.

4 Notes

1. Using an NH_3 analyzer instead of the filter papers for NH_3 detections might be effective for laboratory experiments.
2. Materials adsorbing water will also adsorb NH_3. A chamber made of coated polycarbonate might be better than that made of acrylic plastic because of lower water adsorption on polycarbonate surfaces [16].
3. Installing a small fan into a chamber would be better for precise estimations of the gas exchange parameters such as total leaf resistance to water vapor (r_{tot}) and the molar fractions of NH_3 surrounding the leaves (n_a). However, NH_3 appears to adhere on materials that are made of metals, particularly when air inside the chambers is humid. Using a metal-made part inside the chambers would not be recommended [16].
4. For example, 400–550 mM sorbitol for leaf blades of *Oryza sativa* [6] and 400 mM mannitol for *Brassica napus* leaves [13].

5. The OPA solution without 2-ME becomes oxidized and ineffective.
6. Formate solution is used for stabilizing the NH_4^+ concentrations in the solutions [15].
7. For example, 100–120 min for leaf blades of *Oryza sativa*.
8. Supplying NH_3 from a NH_3 cylinder into the air-inlets of the chambers or into GC may control the chamber NH_3 concentrations within a precise manner.
9. It is best to check if cytoplasmic contaminations occur according to the method described by [13].
10. Approximately 80 μL of the apoplast solutions should be obtained from the lamina of about 10 cm in length for *Oryza sativa*.
11. The apoplast solutions are diluted with the sorbitol solutions. The NH_4^+ and H^+ concentrations of the apoplast solutions might be influenced by those contained in the sorbitol solutions. On the other hand, it is reported that the leaf apoplastic tissues have a buffer capacity to maintain stable NH_4^+ and H^+ concentrations in the apoplast solutions against the artificial changes in NH_4^+ concentrations of the sorbitol solutions [13].
12. Also refer to the equations provided by Mattsson et al. [17].

References

1. Morgan JA, Parton WJ (1989) Characteristics of ammonia volatilization from spring wheat. Crop Sci 29:726–731
2. Weiland RT, Stutte CA (1985) Oxygen influence on foliar nitrogen loss from soyabean and sorghum plants. Ann Bot 55:279–282
3. Husted S, Mattsson M, Möllers C, Wallbraun M, Schjoerring JK (2002) Photorespiratory NH_4^+ production in leaves of wild-type and glutamine synthetase 2 antisense oilseed rape. Plant Physiol 130(2):989–998
4. Kumagai E, Araki T, Hamaoka N, Ueno O (2011) Ammonia emission from rice leaves in relation to photorespiration and genotypic differences in glutamine synthetase activity. Ann Bot 108(7):1381–1386
5. Wang L, Pedas P, Eriksson D, Schjoerring JK (2013) Elevated atmospheric CO_2 decreases the ammonia compensation point of barley plants. J Exp Bot 64(10):2713–2724
6. Miyazawa S-I, Hayashi K, Nakamura H, Hasegawa T, Miyao M (2014) Elevated CO_2 decreases the photorespiratory NH_3 production but does not decrease the NH_3 compensation point in rice leaves. Plant Cell Physiol 55(9):1582–1591
7. Farquhar GD, Firth PM, Wetselaar R, Weir B (1980) On the gaseous exchange of ammonia between leaves and the environment: determination of the ammonia compensation point. Plant Physiol 66:710–714
8. Farquhar GD, Wetselaar R, Weir B (1983) Gaseous nitrogen losses from plants. In: Freney JR, Simpson JR (eds) Gaseous loss of nitrogen from plant-soil systems. Springer, Dordrecht, pp 159–180
9. Hayashi K, Hiradate S, Ishikawa S, Nouchi I (2008) Ammonia exchange between rice leaf blades and the atmosphere: effect of broadcast urea and changes in xylem sap and leaf apoplastic ammonium concentrations. Soil Sci Plant Nutr 54:807–818
10. https://innovation.nilu.no/wp-content/uploads/sites/14/2020/01/NILU-Filter_holder_system.pdf
11. von Caemmerer S, Farquhar GD (1981) Some relationships between the biochemistry of photosynthesis and the gas exchange of leaves. Planta 153:376–387
12. Miyazawa S-I, Nishiguchi M, Futamura N, Yukawa T, Miyao M et al (2018) Low

assimilation efficiency of photorespiratory ammonia in conifer leaves. J Plant Res 131(5): 789–802

13. Husted S, Schjoerring JK (1995) Apoplastic pH and ammonium concentration in leaves of *Brassica napus* L. Plant Physiol 109:1453–1460
14. Goyal SS, Rains DW, Huffaker RC (1988) Determination of ammonium ion by fluorometry or spectrophotometry after on-line derivatization with *o*-phthalaldehyde. Anal Chem 60: 175–179
15. Husted S, Hebbern CA, Mattsson M, Schjoerring JK (2000) A critical experimental evaluation of methods for determination of NH_4^+ in plant tissue, xylem sap and apoplastic fluid. Physiol Plant 109:167–179
16. Schjoerring JK, Husted S (1997) Measurement of ammonia gas emission from plants. In: Linskens HF, Jackson JF (eds) Plant volatile analysis. Springer, Berlin, pp 73–95
17. Mattsson M, Herrmann B, David M, Loubet B, Riedo M et al (2009) Temporal variability in bioassays of the stomatal ammonia compensation point in relation to plant and soil nitrogen parameters in intensively managed grassland. Biogeosciences 6:171–179

Chapter 12

Using Gas Exchange to Study CO_2 Release During Photosynthesis with Steady- and Nonsteady-State Approaches

Stephanie C. Schmiege, Berkley J. Walker, and Thomas D. Sharkey

Abstract

Measures of respiration in the light and C_i^* are crucial to the modeling of photorespiration and photosynthesis. This chapter provides background on the equations used to model C_3 photosynthesis and the history of the incorporation of the effects of rubisco oxygenation into these models. It then describes three methods used to determine two key parameters necessary to incorporate photorespiratory effects into C_3 photosynthesis models: respiration in the light (R_L) and C_i^*. These methods include the Laisk, Yin, and isotopic methods. For the Laisk method, we also introduce a new rapid measurement technique.

Key words Respiration in the light, Laisk method, Kok method, Yin method, Isotopes, Γ_*, C_i^*

1 Introduction

Net gas exchange measurements have been essential for linking photorespiration to carbon assimilation. Evidence for photorespiration and other respiration in the light has been examined for the last 100 years. For example, Warburg [1] used gas exchange methods to discover that oxygen inhibited photosynthesis [2]. Decker [3] showed that CO_2 evolution immediately after imposing darkness was significantly greater than a somewhat stable rate of CO_2 evolution reached several minutes after imposing darkness. (This Post-Illumination Burst is explored further in this book.) The interpretation was that there is one or more processes that release (or "respire") CO_2 and that are stimulated by light. This was variously called photorespiration, for example, Rabinowitch [4] and/or light respiration (R_L). The possibility of light-stimulated CO_2 release was called "a nightmare oppressing all who are concerned with the exact measurement of photosynthesis" by Rabinowitch [4]. Given the importance of gas exchange to the discovery of photorespiration and the importance of photorespiration to

Berkley J. Walker (ed.), *Photorespiration: Methods and Protocols*, Methods in Molecular Biology, vol. 2792,
https://doi.org/10.1007/978-1-0716-3802-6_12,

predicting gas exchange of photosynthesis, we present here a summary of the discoveries of photorespiration by gas exchange and how these discoveries informed the subsequent equations used to model it. This historical perspective is helpful to illustrate the various assumptions integrated into commonly used forms of these models. We then present methods for exploring photorespiration using these models and gas exchange methods based on steady- and nonsteady-state assumptions. In this chapter, we will focus specifically on C_3 photosynthesis.

A significant part of CO_2 release during photosynthesis was found to be associated with, but not necessarily coming directly from, glycolate metabolism in peroxisomes [5]. The source of the glycolate was found to be dependent on rubisco [6]. The metabolism that involves CO_2 released during metabolism of glycolate produced by oxygenation of ribulose 1,5-bisphosphate (RuBP) is now the definition of photorespiration, while other possible CO_2-releasing processes occurring in the light were collectively called day respiration [7]. This nomenclature was the result of the initial belief that this CO_2 release comes from mitochondrial reactions and is the same as respiration in the dark [8] and so-called R_d. To allow that there may be other sources of CO_2 release during photosynthesis that are important to understanding gas exchange behavior, this daytime "dark respiration" was rebranded as "day respiration" so that the abbreviation could be retained. Here, we will define respiration in the light (R_L) as respiration (i.e., CO_2 release) in the light that is not photorespiration. While this could involve mitochondrial metabolism, isotope studies have consistently shown that there is very little activity of the tricarboxylic acid cycle in the light [9–11].

The discovery of rubisco-catalyzed RuBP oxygenation as the initial event in photorespiration opened the door to a quantitative description of the effect of oxygenation on net photosynthesis [12]. Thus,

$$A = v_c - t\tilde{n}v_o \tag{1}$$

where A is net CO_2 assimilation (what we measure in a gas exchange system), v_o and v_c are the velocities of the oxygenase and carboxylase, respectively, and t is the proportion of carbon released as CO_2 during metabolism resulting from one oxygenation event. The actual velocities are denoted by lower case v's as is the norm in enzymology, to distinguish from theoretical maximum velocities, V_{max}. The value of t has been taken to be 0.5 [13] based on the glycolate metabolism pathway proposed by Tolbert [14]. The relationship between v_o and v_c, that is, the ratio of oxygenation to carboxylation, is labeled Φ and given by

$$\frac{v_o}{v_c} = \frac{V_o K_c}{V_c K_o} \tilde{n} \frac{O}{C} = \Phi \tag{2}$$

where V_o and V_c are V_{max} for the oxygenase and carboxylase activities, K_c and K_o are the Michaelis constants for carboxylation and oxygenation, and O and C are the partial pressures or concentrations of oxygen and CO_2. From Eq. 2, it is clear that the ratio of oxygenation to carboxylation is linearly dependent on oxygen and inversely dependent on CO_2. Keck and Ogren [13] used Eqs. 1 and 2 to derive the CO_2 partial pressure (or concentration as long as the K_m's are in the same units) at which CO_2 assimilation by carboxylation is equal to CO_2 release by oxygenation, the compensation point, Γ.

$$\Gamma = \frac{tV_oK_cO}{V_cK_o}. \tag{3}$$

Farquhar et al. [8] pointed out that to apply the above equations to gas exchange measurements of photosynthesis and related photorespiration, it was necessary to account for R_L. Thus, Eq. 1 becomes

$$A = v_c - t\tilde{n}v_o - R_L. \tag{4}$$

The rubisco compensation point as defined in Eq. 3 will occur when gas exchange is showing a CO_2 release equal to R_L. Farquhar et al. [8] renamed Γ as defined initially by Eq. 3 as Γ_* since it is the rubisco compensation point, not the point where net leaf CO_2 exchange is zero [*see* 15 for full derivations]. Thus, the CO_2 partial pressure at which CO_2 exchange is independent of light intensity is Γ_* (but see below regarding $C_i{}^*$). And, at Γ_*, the measured CO_2 exchange is a measure of R_L.

Using Φ as defined in Eq. 2 and combining it with Eq. 4 gives

$$A = v_c(1 - t\Phi) - R_L. \tag{5}$$

Equations 2 and 3 can then be used to derive that

$$\frac{v_o}{v_c} = \frac{1}{t}\frac{\Gamma_*}{C} \text{ typically shown to be } \frac{2\tilde{n}\Gamma_*}{C} \text{ if } t = 0.5. \tag{6}$$

Laisk [16] pointed out that at Γ_*, A will be independent of light and so a series of CO_2 response curves at limiting light levels will cross over at Γ_*. This makes it possible to use gas exchange measurements to determine a parameter that combines many rubisco characteristics, or in other words, provides a powerful validation of the above theory when they do cross over in vivo at a value predicted by rubisco kinetics in vitro. However, Γ_* depends on t, and so if t is variable, Γ_* will be variable.

However, the point at which various CO_2 response curves cross over will not be Γ_* if there are other sources of CO_2 release or if there is diffusion resistance between the intercellular air spaces and rubisco. This is because the diffusion resistance encountered by the net flux of CO_2 across the mesophyll will cause the apparent Γ_* to be at a lower CO_2 partial pressure than the true Γ_*. This apparent Γ_* is called C_i* [17]. The relationship between Γ_* and C_i* is

$$\Gamma_* = C_i^* + \frac{R_L}{g_m} \tag{7}$$

where g_m is the diffusion conductance between the intercellular air spaces and rubisco. Notably, if there are large diffusive barriers between the mitochondrial release of CO_2 and rubisco, Eq. 7 is not as valid, and an additional theoretical framework involving multiple resistances is needed [18].

Gas exchange measurements can be used to estimate both C_i* and R_L, which, as demonstrated through the theoretical equations above, are important parameters to estimates of rubisco oxygenation and thus photorespiration. Here, we outline several gas exchange methods used to estimate C_i* and R_L, any recent advances in these methods, and any important considerations when using these methods. Both C_i* and R_L can be measured using the Laisk method [Laisk [19] as described in English in Laisk [16]], whereas R_L can also be measured by several other techniques including the Kok [20] and Yin [21, 22] methods, and by using isotopes [23, 24]. Materials for these methods are listed below followed by the protocols themselves. Finally, we present additional insight into isotopic methods gained from metabolic flux analysis studies.

2 Materials

2.1 Laisk, Kok, and Yin Materials

1. Infra-Red Gas analyzer (IRGA)-based gas exchange system. The system should have the capability to measure both gas exchange and chlorophyll fluorescence simultaneously for the Yin method.

2.2 Isotopic Materials

1. IRGA.
2. Gas tanks: 99% $^{13}CO_2$, 99% $^{12}CO_2$, N_2 and O_2.
3. 5 mass flow controllers.
4. Swagelok T-joint attached to the back of the IRGA measurement head to facilitate easy switching from $^{12}CO_2$ to $^{13}CO_2$.
5. Bev-A-line IV tubing (more gas-tight than Teflon tubing).
6. Bubbler to humidify airstream.
7. Tunable diode laser (TDL) or equivalent method capable of identifying $^{12}CO_2$ emission from a leaf in a $^{13}CO_2$ background.

3 Methods

This chapter does not provide detailed instructions on best practices when using gas exchange systems as many such guides already exist in the literature (*see* e.g., [25]); nonetheless, we do feel it is important to have a basic understanding of the measures and units used to describe the proportions of a gas in air. As such, we have included a basic primer on this topic in Appendix 1.

3.1 Laisk Method for Estimating C_i* and R_L

The Laisk method [16, 19] estimates C_i* and R_L by collecting at least two but in practice typically three to five photosynthetic CO_2 response curves at different light intensities such that the curves intersect at a single point where the *x* and *y* coordinates are equal to C_i* and R_L, respectively (Fig. 1). While the expectation is that all the CO_2 response curves should cross over at the same point, it has frequently been documented that the cross-over points can differ among pairs of curves [18]. For this reason, several methods have been developed to identify a common crossover point from the curves including averaging the values obtained from the intersection of each pair of curves and the slope-intercept regression method [26, 27]. The CO_2 response curves used to identify this cross-over point are collected either by steady-state gas exchange techniques, or by employing the new dynamic assimilation technique (DAT; Fig. 1) [28]. The DAT technique significantly reduces the time required to collect the CO_2 response curves allowing for higher throughput and leaving less time for the physiology of the leaf to change in response to holding them at or below the compensation point. Both steady-state and DAT techniques provide comparable estimates of R_L, although estimates of C_i* may vary slightly between the two techniques [29]. At this stage, it is not fully clear why there are slight differences between techniques in determining C_i*, but it is possible that these small shifts may be due to slight changes in g_m or in glycine export from the photorespiratory pathway [29].

3.1.1 Steady-State Laisk Protocol

1. Identify at least three light intensities (*see* **Notes 1** and **2**) that provide evenly spaced differences in the initial slopes of the CO_2 response curves.
2. Identify CO_2 concentrations that span the linear portion of the CO_2 response curve but minimize the amount of time spent at very low CO_2 concentrations (*see* **Note 3**). Stay at each CO_2 concentration for at least 30 s but no more than 120 s. Point matching should be employed before each measurement in IRGA's with two separate detectors (*see* **Note 4**).

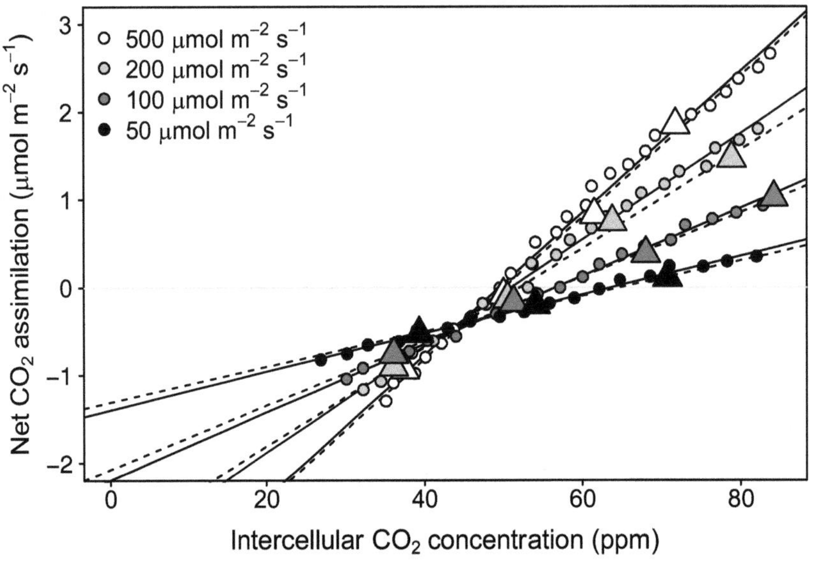

Fig. 1 Example comparison of the data collected from DAT and steady-state Laisk methods and corresponding linear fits (DAT, circles and dashed lines; steady-state, triangles and solid lines) from *Betula papyrifera*

3. Run CO_2 response curves at each light intensity, returning to 420 ppm between each curve until all light levels have been completed.

3.1.2 DAT Laisk Protocol (Using the LI-6800)

The LI-6800 is currently the only IRGA capable of running DAT curves, so specific instructions are provided for this particular instrument.

1. Set up DAT on the LI-6800 (*see* **Note 5**). We have found that this technique works best with LI-6800s that have the most recent processor installed. This processor decreases lagging in the measurements during the CO_2 ramps.
   (a) Enable dynamic equations.
   (b) Test dynamic tuning using an empty chamber and your chosen flow rate (usually between 300 and 600 μmol s^{-1}).
   (c) Set up range matching.
2. Light intensities for each of the curves should be identified as above.
3. At the first light intensity, ramp the reference CO_2 concentration from high to low such that you collect the approximately linear portion of a traditional A/C_i curve (*see* **Note 6**).
4. Return to 420 ppm (ambient atmospheric CO_2 concentration) before running the curve at the next light level to minimize time spent at low CO_2 concentrations.

5. Point match before running the next curve, especially if the LI-COR range match was set up before the IRGA was fully warmed up (within the first hour of starting up the LI-COR).
6. Repeat steps two through four until curves at all light levels have been completed.

3.1.3 Laisk Curve Analysis Protocol

This method of extracting C_i* and R_L from the Laisk curves uses the slope-intercept regression method [26, 27].

1. If using DAT-collected Laisk curves, remove the first five or so points from each curve as they show the initial adjustment to the CO_2 ramp and are not linear.
2. Visually assess the linearity of the data you have collected. Subset to datapoints in the linear portion. We have found these to be the points below 85–100 ppm.
3. Fit linear regressions to the CO_2 response curves at each of the light intensities.
4. Extract the slope and intercept of the linear regressions (*see* **Note 7**).
5. Fit a linear regression to the slopes and intercepts from **step 4** with the slopes on the *x*-axis and the intercepts on the *y*-axis. The slope and intercept of this line provide estimates of C_i* and R_L, respectively.

3.2 Kok and Yin Methods for Estimating R_L

In addition to the Laisk method, R_L can also be estimated via the Kok or Yin method. In these methods, R_L is estimated by collecting a photosynthetic light response curve with particular attention to low light intensities around the light compensation point. The Kok method derives its name from Bessel Kok, who discovered a subtle shift in the response of photosynthesis to light intensity around the light compensation point, now called the "Kok effect" [20]. The point where this shift occurs has been called the breakpoint. Biologically, it has been interpreted as the point where leaf mitochondrial respiration is suppressed by light. Consequently, if a linear regression is fit to the points above the breakpoint, the *y*-intercept will provide an estimate of R_L. In contrast, if a linear regression is fit to the points below the breakpoint, an estimate of respiration in the dark (R_D) is gained instead. An important update for this method requires accounting for the fact that internal CO_2 concentrations (C_i) increase as light intensity decreases [30]. The higher C_i at low light levels suppresses photorespiration relative to carboxylation resulting in higher measured photosynthetic rates in the linear portion of the curve [31]. The result is a lower slope in the linear relationship through the observed data than would be observed if the C_i were constant, resulting in an underestimation of R_L. Correcting for changes in C_i should be undertaken and methods for this are fully described in Kirschbaum, Farquhar [32], and Ayub et al. [30].

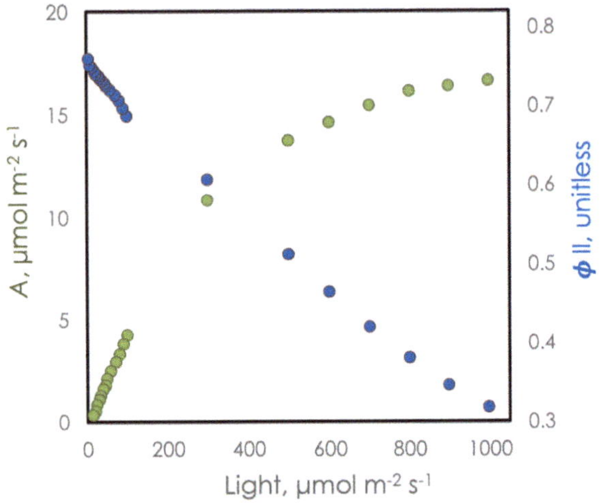

Fig. 2 Response of photosynthesis (*A*) and Φ_{II} to changes in light

More recently, an additional modification to the Kok method has been proposed. The Kok method assumes that photosystem II electron transport efficiency (Φ_{II}) is constant across all light levels used in the analysis; however, Φ_{II} declines at high light levels (Fig. 2). This has led to the advent of the Yin method, developed by Xinyou Yin [21, 22], which incorporates this decline in Φ_{II}. Φ_{II} is estimated by taking simultaneous measurements of chlorophyll fluorescence during gas exchange. At each light intensity of the photosynthetic light response curve, Φ_{II} is calculated as $1 - F_s/F_m'$. Then, Φ_{II} is incorporated into the traditional light response curve by plotting photosynthesis as a function of $I_{inc}\ \Phi_2/4$ where I_{inc} is the incident irradiance. R_L is estimated from this modified light response curve by extracting the intercept from the linear regression of photosynthesis to $I_{inc}\ \Phi_{II}/4$ in the lower portion of the response curve. These estimates of R_L are often slightly larger than estimates from the Kok method, but comparable to estimates acquired via the Laisk method [22].

3.2.1 Kok and Yin Protocol

1. Data for Kok and Yin methods are acquired using steady-state gas exchange techniques. Data for the Kok method can be collected using an IRGA with precise light intensity control. Data for the Yin method must be collected with an IRGA capable of acquiring simultaneous chlorophyll fluorescence and gas exchange measurements.
2. For the Kok and Yin methods, a larger number of datapoints should be acquired at lower light intensities (*see* **Note 8**).

3.3 Isotopic Methods for Estimating R_L

In addition to gas exchange methods, there are also isotopic methods that have been used to estimate R_L. Although these methods are not easily employed in field settings, limiting their application, they have been lauded for their ability to measure R_L under high light and photosynthetic conditions. The fact that Laisk, Kok, and Yin methods require altering CO_2 concentrations and/or light conditions at conditions near the CO_2 or light compensation point limits our ability to assess the magnitude of this flux under ambient photosynthetic conditions. One of the first isotopic methods used to assess R_L under high light and high photosynthetic conditions was the method developed by Francesco Loreto [23, 24]. In this technique, the leaf is rapidly transitioned to 99.9% $^{13}CO_2$ environment and the $^{12}CO_2$ emission from the leaf (measured using an IRGA with reduced sensitivity to $^{13}CO_2$) is measured as an estimate of R_L. Theoretically, this method provides accurate estimates of R_L because CO_2 released by R_L comes from older stored pools of carbon that will not be labeled by exposure to $^{13}CO_2$. The advantage of this method is that R_L can be assessed under a variety of different light conditions and CO_2 concentrations. Nevertheless, there are several caveats that are important to be aware of when using this method.

The first of these is that the intermediates of the Calvin Benson cycle label quickly during the first 5–10 min but are not fully labeled for many hours [33]. We know that the Calvin Benson cycle intermediates are 80–90% labeled after 20 min and can take this degree of label into account when estimating R_L. However, because the Calvin Benson cycle is incompletely labeled, we can infer that photorespiration also remains incompletely labeled to some degree. Up until now, there have been few estimates of the degree of label in photorespiration, making it challenging to ascertain whether the total measured $^{12}CO_2$ efflux from the leaf is due to R_L or the incomplete labeling of photorespiratory CO_2 release. However, a recent metabolic flux analysis by Xu et al. [10] may provide the information necessary to estimate this (see below).

An additional factor that must be accounted for is the fact that CO_2 released in the cell can be refixed in the Calvin Benson cycle or released into the atmosphere. Any estimates of R_L via this isotopic labeling method must take refixation into account. In fact, a simple mathematical method was used by Loreto et al. [23] in which reassimilated $^{12}CO_2$ was calculated from the ratio of $^{12}C_i$ to $^{13}C_i$ multiplied by the photosynthetic rate. In the second appendix to this chapter, we provide a more complex accounting for refixation that considers the updated understanding of $^{13}CO_2$ labeling time courses, the potential for competitive interactions between the isotopes, and photorespiration effects on the relationship between photosynthesis and the velocity of carboxylation (*see* Appendix 2).

In the method below, we present a modified version of the setup used by Loreto, where instead of a $^{13}CO_2$-insensitive IRGA, we use a TDL tuned to wavelengths that can sensitively detect $^{12}CO_2$ in an enriched $^{13}CO_2$ background with great sensitivity and precision.

$^{13}CO_2$ Labeling Protocol

System setup to prepare the air mixture that will be fed into the gas exchange system

1. Connect the O_2, 99% $^{12}CO_2$ and 99% $^{13}CO_2$ to flow controllers.
2. For the N_2 divide the airstream into two separate tubing paths with separate flow controllers. One path will pass through a bubbler, the other through desiccant. This will allow some control of the humidity in the airstream.
3. The humid and dry N_2 airstreams should then be joined to the O_2 airstream.
4. Install the Swagelok T-joint on the air inlet of the IRGA head.
5. Connect the N_2, O_2 tube, and the $^{13}CO_2$ and $^{12}CO_2$ lines such that turning the four-way switching valve will rapidly add either $^{13}CO_2$ or $^{12}CO_2$ to the N_2, O_2 airstream entering the leaf chamber (*see* Fig. 3 for a flow path diagram of the system setup).
6. Flow controllers should be set to provide 80% N_2, 20% O_2, and 420 ppm of either $^{13}CO_2$ or $^{12}CO_2$.

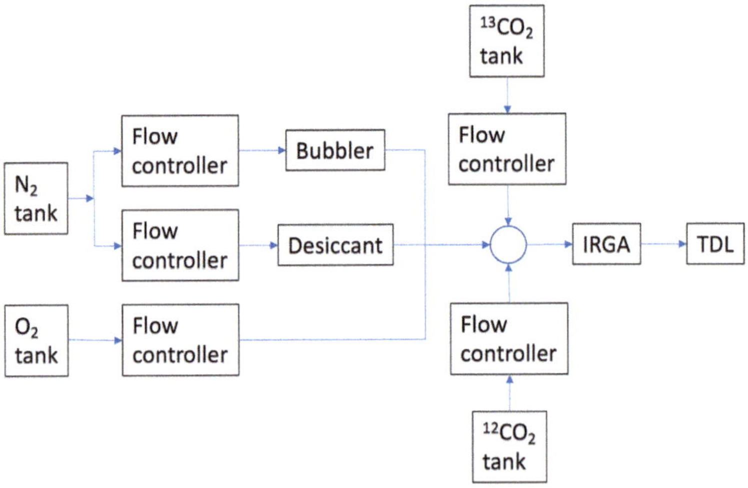

Fig. 3 Flow diagram of the isotope method system setup using the TDL. Flow from the N_2 tank is split into two streams with separate flow controllers. Flow passes through a bubbler or desiccant to control the humidity of the airstream. These separate flows will then join the oxygen tubing path and pass to a four-way switching valve (the circle in the diagram). Flow from the $^{13}CO_2$ tank and the $^{12}CO_2$ tank are connected to the valve such that turning the valve allows the addition of either $^{13}CO_2$ or $^{12}CO_2$ to the N_2 and O_2 airstream. The flow then feeds to the IRGA (and leaf chamber) and then the TDL

7. Connect the chamber (sample) and reference air outlets from the IRGA to a TDL or equivalent system to measure the $^{12}CO_2$ in the exhaust chamber air.

Measurements

1. Let the leaf acclimate in $^{12}CO_2$ until photosynthesis and stomatal conductance are stable (often approximately 20 min). Record photosynthesis, transpiration, and flow rate on the gas exchange system. These will be used later to calculate $^{12}CO_2$ concentrations.
2. After acclimation, switch from $^{12}CO_2$ to $^{13}CO_2$ (Fig. 4). Record the $O_{16}C_{12}O_{16}$ peak of the TDL absorbance spectra for 20 min.

Calculating $^{12}CO_2$ Efflux

1. To calculate the total $^{12}CO_2$ efflux from the $O_{16}C_{12}O_{16}$ peak of the TDL absorbance spectra, you will need to take into account your leaf area and correct your flow rate to account for transpiration as water vapor efflux from the leaf increases the total the flow rate [*see* Appendix 2 in 34].

Insight from Metabolic Flux Analysis

One of the concerns with isotopic methods for estimating R_L is the fact that we do not know how much of the $^{12}CO_2$ efflux from the leaf is due to R_L as opposed to incomplete labeling of photorespiration with $^{13}CO_2$. Metabolic flux analysis provides a unique opportunity to examine the contributions of different CO_2-releasing processes to total $^{12}CO_2$ emission. Xu et al. [10] provide a unique dataset that allows a first approximation of the contributions from the various CO_2-releasing processes in the leaf occurring during photosynthesis. Their dataset includes the degree of label in a variety of different metabolites along with the velocity of CO_2 release from the enzymes catalyzing the processes. From these data, we can calculate the total ^{12}C emission from a leaf as the sum of the CO_2 releasing fluxes multiplied by the degree of ^{12}C label remaining in the metabolites after 30 min in a 99% pure $^{13}CO_2$ environment if we assume that the enrichment of these metabolites is the same during a $^{13}CO_2$ experiment measured with the TDL. Thus, contributing pathways could include the glucose 6-phosphate (G6P) shunt, fatty acid synthesis, photorespiration, and the tricarboxylic acid (TCA) cycle such that:

$$
\begin{aligned}
{}^{12}C = & (1 * R_{\mathrm{UDPG}} * v_{\mathrm{6PGD}}) + (1 * R_{\mathrm{PEP}} * v_{\mathrm{PDH.c}}) \\
& + (0.5 * R_{\mathrm{RUBP}} * v_{\mathrm{o}}) + (1 * R_{\mathrm{PEP}} * v_{\mathrm{PDH.m}}) \\
& + (1 * R_{\mathrm{ICI}} * v_{\mathrm{IDH}}) + (1 * R_{\mathrm{ICI}} * v_{\mathrm{KGDH}})
\end{aligned} \tag{12}
$$

where ^{12}C is the total ^{12}C emission from the leaf, v is the velocity or rate of flux contributing to the total ^{12}C emission, and R is the %

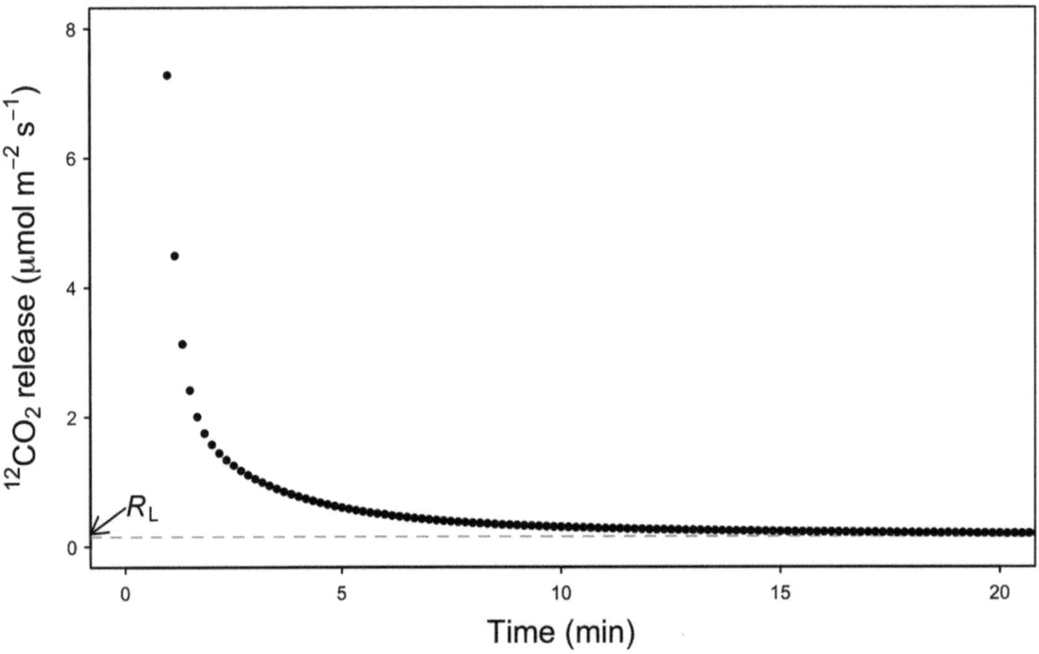

Fig. 4 $^{12}CO_2$ release during the transition from a $^{12}CO_2$ to a $^{13}CO_2$ environment over 20 min calculated from the TDL absorbance spectra. Over 20 min, the Calvin Benson cycle intermediates are labeled with $^{13}CO_2$ and the $^{12}CO_2$ release from the leaf decreases. At 20 min the Calvin Benson cycle is 80–90% labeled and the remaining $^{12}CO_2$ release can be attributed to R_L (the *y*-intercept; but see chapter for important caveats regarding the incomplete labeling of photorespiration and possible impacts of R_L estimates from this method)

^{12}C label remaining in the metabolites. UDPG, UDP glucose; 6PGD, 6 phosphogluconate dehydrogenase; PEP, phosphoenolpyruvate; PDH.c, chloroplastic pyruvate dehydrogenase; RUBP, ribulose-1,5-bisphosphate, v_o, velocity of oxygenation; PYR.m, mitochondrial pyruvate; PDH.m, mitochondrial pyruvate dehydrogenase; ICI, isocitrate; IDH, isocitrate dehydrogenase; KGDH, α-ketoglutarate dehydrogenase.

Note that in some cases the nearest upstream metabolite was not available and so nearby representative metabolites were chosen instead. Thus, in the G6P shunt, we have used % ^{12}C release from UDPG instead of 6-phosphogluconate (6PG) as 6PG is very hard to estimate. In fatty acid synthesis, we have used PEP instead of chloroplastic pyruvate (PYR.c) as evidence from work on isoprene by Sharkey et al. [33] indicates that sources of carbon for the methyl erythritol pathway, glyceraldehyde 3-phosphate, and pyruvate are labeled to a similar degree as CBC intermediates. In photorespiration, we have used RUBP instead of glycine because glycine can also be stored in the vacuole making it challenging to differentiate between slow and fast pools of this metabolite. The % label in glycine would need to be the total of both slow and fast pools. Thus, the % ^{12}C label in glycine would be most accurately characterized according to the following equation:

$$R_{GLY} = (v_o * R_{RUBP} + v_{GLY.v} * R_{GLY.v})/(v_o + v_{GLY.v}) \quad (13)$$

where the total label in glycine includes the rate of exchange with, and ^{12}C label in, the slow, vacuolar pool of glycine (GLY.v). As we do not have estimates of GLY.v, we have used RUBP to set the degree of label in glycine. Finally, in the TCA cycle CO_2 releasing reactions, we have used the label in PEP instead of in mitochondrial pyruvate because there is large variability in the pool of pyruvate and we have used the label in ICI instead of α-ketoglutarate as we have no estimate of the % ^{12}C label in this α-ketoglutarate.

It is worth noting here that there is a possible additional CO_2-releasing pathway in which malate is decarboxylated by malic enzyme to form pyruvate. This reaction has not previously been considered in flux analysis studies. Consequently, we do not know the velocity of this reaction compared to other CO_2-releasing reactions and have therefore excluded it from this current assessment. Future studies could examine this further.

By using the % ^{12}C label from the identified metabolites from Xu et al. [10], and converting the velocities from μmol metabolite g^{-1}FW h^{-1} to μmol m^{-2} s^{-1} using the ratio of fresh weight to area of 550 g m^{-2}, we have calculated the contribution to $^{12}CO_2$ release from each of the processes (Table 1). From these calculations, we estimate that the processes usually considered to contribute to R_L (the G6P shunt, fatty acid synthesis, and the TCA cycle) release a total of 0.374 μmol m^{-2} s^{-1} CO_2, while photorespiration releases 0.245 μmol m^{-2} s^{-1} CO_2. Thus, we can see that photorespiration comprises a large fraction of the total $^{12}CO_2$ release as measured using isotopic methods, and accurately accounting for this photorespiratory contribution is critical to accurate estimations of R_L via this technique. Not only is this new accounting important for

Table 1
Remaining % ^{12}C label in metabolites at 30 min (means, $n = 3$), and velocities of ^{12}C emission [10] used to calculate $^{12}CO_2$ release from the leaf

	% ^{12}C label		Velocity (μmol m^{-2} s^{-1})		$^{12}CO_2$ release (μmol m^{-2} s^{-1})
G6P Shunt	UDPG	0.214	v_{6PGD}	1.069	0.229
Fatty acid synthesis	PEP	0.11	$v_{PDH.c}$	0.061	0.007
Photorespiration	RUBP	0.063	v_o	7.792	0.245
TCA cycle	PEP	0.11	$v_{PDH.m}$	0.141	0.015
	ICI	0.873	v_{IDH}	0.141	0.123
	ICI	0.873	v_{KGDH}	0.000	0.000
Total $^{12}CO_2$ release					0.619

isotopic methods, but it highlights that photorespiration contributes a large proportion to the total CO_2 release during photosynthetic daylight hours.

4 Notes

1. When setting the light intensity on a LI-COR gas exchange instrument such as the LI-6800, we set the proportion of red light and blue light reaching the leaf is 50:50, which might be better for keeping the stomata open during multiple rounds of variation in light and CO_2 [35].
2. An initial light response curve can be used to select light intensities that will give an even spread of slopes during the Laisk measurement.
3. Generally, it is not recommended to go below 25 ppm when taking steady-state gas exchange measurements. We have used the following CO_2 concentrations with good results: 150, 100, 75, 50, and 25 ppm.
4. Some gas exchange systems are not capable of measuring at CO_2 levels below the cross-over point. It is possible to use projections from the higher CO_2 concentrations, but this is not optimal.
5. For all Laisk measurements, if using a gas exchange system with fluorescence capabilities, make sure that fluorescence is turned off.
6. We have found that starting at 150 ppm and ramping to 0 ppm at a rate of 50 ppm min^{-1} works well.
7. There is a spreadsheet available in the supplemental information of [27] to perform the analysis according to the slope-intercept regression method.
8. We recommend collecting the photosynthetic and fluorescence measurements at the following light intensities: 100, 75, 50, 45, 40, 35, 30, 25, 20, 15, 10, 5 $\mu mol_{photons}\ m^{-2}\ s^{-1}$. These measurements can be combined with a full light response curve if desired.

Appendix 1: A Primer on Gas Exchange – Measures of the Proportion of a Gas in Air

There are several aspects of using gas exchange to study photorespiration that can be confusing initially. To start, how should one describe the amount of gas being used? Gases dissolve into liquids in proportion to their partial pressure (Henry's law). In ideal gases, the total pressure is the sum of the pressures that would be exerted by each component (Dalton's law). A common unit of pressure is the standard atmosphere (at sea level). The SI unit for pressure is the Pascal (Pa), one standard atmosphere is 101.3 kPa. People use bar, which is convenient because 1 bar is 1.013 atmospheres. The atmosphere with a sea-level pressure of 101.3 kPa total pressure would have about 78 kPa nitrogen, 21 kPa oxygen, 1 kPa argon, 42 Pa CO_2, and 0 to ~4 kPa water vapor.

For photorespiration studies, we want to know the availability of CO_2 and oxygen. According to Henry's law, CO_2 dissolves into the water-saturated cell walls inside a leaf according to its partial pressure. To illustrate, at the top of a mountain the ratio of oxygen partial pressure to total pressure is the same as at sea level, but the total pressure is less and so the partial pressure of oxygen is less, making it hard to breathe.

Most often people express CO_2 in parts per million (and oxygen in %). These are unitless ratios (% and PPM are not units, they are used when units cancel). A very useful fact is that the partial pressure of a gas divided by the total pressure is the same as the partial volume of a gas divided by the total volume or the number of moles of the gas divided by the total number of moles of all gases present. This is a mole fraction and denoted χ. Since there are only 0.00042 moles of CO_2 in a mole of air, we express this as moles of CO_2 per million moles of air, ppm. This is different from the ppm used in fertilizer studies. In that usage, 1 ppm is 1 mg per liter. Since milligrams and liters are not the same, use of ppm in this context is often frowned upon, but in gases, expressing mole fraction is defensible. However, the criticism of mg per liter as ppm spilled into gas studies so now to avoid saying ppm we use μL L^{-1} or, because lowercase l is confused with the numeral one, μL L^{-1}. Others use μmol mol^{-1} and also μbar bar^{-1} or μPa Pa^{-1}. They are all mole fraction and identical (Table 2).

$$\text{mole fraction}, \chi = \frac{\text{mol}}{\text{mol}} = \frac{\text{Pa}}{\text{Pa}} = \frac{\text{L}}{\text{L}} \qquad \text{(A1.1)}$$

So, when should mole fraction be used and when should partial pressure be used? When communicating about how CO_2 affects photosynthesis, it is best to use partial pressure. That way, the effective CO_2 availability is the same regardless of total pressure. A CO_2 response curve reported in partial pressure will be the same at sea level (101.3 kPa atmospheric pressure) and in Denver Colorado, USA (84 kPa). If you report in mole fraction, then the effective CO_2 availability for photosynthesis in Denver will be only 83% of what was available at sea level. On the other hand, mole fraction is often the more convenient measure in the lab. Most mass flow meters report the molar flow of a gas. If you mix two gas streams, you will know the ratio of the molar flux of each. If you mix them at high pressure and reduce the pressure, the mole fraction will stay the same while the partial pressure will change. This is especially

Table 2
Expressing the amount of water vapor in air at 25, 30, and 35 °C

Temperature, °C	Vapor pressure, kPa	Relative humidity, %	Partial pressure, kPa	Mole fraction, %	Dew point, °C
Constant relative humidity of 60%					
25	3.17	60	1.90	1.88	16.7
30	4.24	60	2.54	2.51	21.4
35	5.63	60	3.38	3.33	27.8
Constant vapor pressure difference of 1.5 kPa					
25	3.17	53	1.67	1.64	14.7
30	4.24	65	2.74	2.37	20.2
35	5.63	73	4.13	4.08	29.3

The water vapor pressure above liquid water at the indicated temperature is in the second column. The remainder of the columns are for a relative humidity of 60%, a common target humidity used in gas exchange studies. The mole fraction assumes an atmospheric pressure of 101.3 kPa. From this table, it is clear that in order to make gas exchange measurements at 35 °C to examine the effect of temperature on photorespiration, it would be necessary to do the experiment in a warm greenhouse or growth chamber or accept less than 60% relative humidity (the alternative of risking condensation in the gas exchange system is not advised). Table 2 shows that it is difficult to set humidity to be constant at higher temperatures. If you use relative humidity, then the absolute humidity (partial pressure) will vary. If you set the vapor pressure difference between the leaf and air constant, then relative humidity will be different

applicable to isotope studies. We routinely start with a pressure vessel with a known amount of $^{13}CO_2$ or $^{14}CO_2$ and then pressurize the tanks. In this case, partial pressure can be ignored, just the molar ratios need to be considered.

Mole fraction and partial pressure issues also apply to water vapor but there is another consideration for water vapor, the dew point. This is the temperature at which humid air has as much water vapor as possible. Any colder and condensation will occur. Condensation is an all-too-common disaster in gas exchange systems. So, in addition to mole fraction and partial pressure, there are two additional ways to describe how much water vapor is in the air. The first is the dew point of the air, that is the temperature at which dew (condensation) would occur if the air comes to that temperature, regardless of the current air temperature. The partial pressure of water vapor above liquid water, often denoted e_0, is a function of absolute temperature and appears exponential. Thus, an empirical equation to determine the partial pressure above liquid water (e_0) in kPa, where T is temperature in degrees Celsius [36] is

$$e_0 = 0.61121^{\left(\left(18.678 - \frac{T}{234.5}\right)\tilde{n}\left(\frac{T}{257.14+T}\right)\right)} \quad \text{(A1.2)}$$

In gas exchange, we estimate the partial pressure of water vapor in the airspaces inside the leaf by knowing the leaf temperature and looking up in a table (or using an empirical equation) to determine the partial pressure of water vapor for pure water at that temperature. On the other hand, relative humidity is very often used to describe the amount of water vapor in air. This is the partial pressure of water vapor divided by the partial pressure that the air at that temperature could hold before condensation would occur. Table 2 shows how these measures of water vapor are related at three temperatures.

Appendix 2: Accounting for Refixation in Isotopic Methods

Loreto et al. (2001) originally accounted for refixation according to the following equation:

$$R_{LR} = {}^{12}C_i/{}^{13}C_i \cdot A \tag{A2.1}$$

where R_{LR} is released $^{12}CO_2$ that is reassimilated, $^{12}C_i$ is calculated below, and $^{13}C_i$ is calculated from gas exchange. A is net CO_2 assimilation.

$$^{12}C_i = R_L/g_s \tag{A2.2}$$

where R_L is the rate of respiration in the light and g_s is stomatal conductance to CO_2.

Given our updated understanding of $^{13}CO_2$ labeling, it may be necessary to elaborate on the original equations. Both oxygen and $^{13}CO_2$ will be competitive inhibitors of $^{12}CO_2$ fixation and vice versa, thus, the ratio of $^{12}CO_2$ carbon fixation to $^{13}CO_2$ carbon fixation can be described as:

$$\frac{^{12}v_C = \cancel{V_{Cmax}} \cdot {}^{12}C/\left({}^{12}C + \left(K_C \cdot \left(1 + {}^{O}\!/_{K_O} + {}^{13}C/K_C\right)\right)\right)}{^{13}v_C = 0.97 \cdot \cancel{V_{Cmax}} \cdot {}^{13}C/\left({}^{13}C + \left(K_C \cdot \left(1 + {}^{O}\!/_{K_O} + {}^{12}C/K_C\right)\right)\right)} \tag{A2.3}$$

(V_{cmax} for $^{13}CO_2$ is 0.97 times that for $^{12}CO_2$). Let us call the ratio of these two equations $^{12/13}R$. In these equations, we calculate the CO_2 concentrations inside the chloroplast by:

$$^{13}C = {}^{13}C_a - \frac{1.6ñAñ0.97}{g_s} - \frac{Añ0.97}{g_m} \tag{A2.4}$$

and

$$^{12}C = {}^{12}C_a + \frac{1.6ñR_L}{g_s} + \frac{R_L}{g_m} \tag{A2.5}$$

Where C_a is the concentration of the indicated isotope of CO_2 in the air outside the leaf and g_m is the mesophyll conductance. These equations assume similar diffusion paths for $^{12}CO_2$ and $^{13}CO_2$ but opposite directions of flux.

If we know total A, we can estimate $^{13}v_C$.

$$A = {}^{T}v_C(1 - 0.5\phi) - R_L \quad (A2.6)$$

where ${}^{T}v_c$ is the total velocity of carboxylation. Then assume ${}^{12}v_C$ is negligible relative to ${}^{13}v_C$ (we estimate 1%), then ${}^{T}v_C \approx {}^{13}v_C$ and so

$${}^{13}v_C = \frac{(A + R_L)}{(1 - 0.5\phi)}. \quad (A2.7)$$

Then the rate of carboxylation of ${}^{12}CO_2$, i.e., refixation, is

$${}^{12}v_C = {}^{12/13}R\tilde{n}{}^{13}v_C\,. \quad (A2.8)$$

References

1. Warburg O (1920) Über die Geschwindigkeit der photochemischen. Kohlensäurezersetzung in lebenden Zellen. II. Biochem Z 103:188–207
2. Kutschera U, Pieruschka R, Farmer S, Berry JA (2020) The Warburg-effects: basic metabolic processes with reference to cancer development and global photosynthesis. Plant Signal Behav 15(7):1776477. https://doi.org/10.1080/15592324.2020.1776477
3. Decker JP (1955) A rapid, postillumination deceleration of respiration in green leaves. Plant Physiol 30(1):82–84
4. Rabinowitch EI (1945) Photosynthesis and related processes. In: Chemistry of photosynthesis, chemosynthesis and related processes in vitro and in vivo, vol I. Interscience Publishers Inc., New York, p 569
5. Tolbert NE, Yamazaki RK (1969) Leaf peroxisomes and their relation to photorespiration and photosynthesis. Ann N Y Acad Sci 168(2 The Nature an):325–341. https://doi.org/10.1111/j.1749-6632.1969.tb43119.x
6. Bowes G, Ogren WL, Hageman RH (1971) Phosphoglycolate production catalyzed by ribulose diphosphate carboxylase. Biochem Biophys Res Commun 45:716–722
7. Tcherkez G, Gauthier P, Buckley TN, Busch FA, Barbour MM, Bruhn D, Heskel MA, Gong XY, Crous KY, Griffin K, Way D, Turnbull M, Adams MA, Atkin OK, Farquhar GD, Cornic G (2017) Leaf day respiration: low CO_2 flux but high significance for metabolism and carbon balance. New Phytol 216(4):986–1001. https://doi.org/10.1111/nph.14816
8. Farquhar GD, von Caemmerer S, Berry JA (1980) A biochemical model of photosynthetic CO_2 assimilation in leaves of C_3 species. Planta 149:78–90
9. Calvin M, Massini P (1952) The path of carbon in photosynthesis. XX. The steady state. Experientia 8(12):445–457
10. Xu Y, Wieloch T, Kaste JAM, Shachar-Hill Y, Sharkey TD (2022) Reimport of carbon from cytosolic and vacuolar sugar pools into the Calvin-Benson cycle explains photosynthesis labeling anomalies. Proc Natl Acad Sci 119(11):e2121531119. https://doi.org/10.1073/pnas.2121531119
11. Xu Y, Fu X, Sharkey TD, Shachar-Hill Y, Walker B (2021) The metabolic origins of non-photorespiratory CO_2 release during photosynthesis: a metabolic flux analysis. Plant Physiol 186:297–314. https://doi.org/10.1093/plphys/kiab076
12. Laing WA, Ogren WL, Hageman RL (1974) Regulation of soybean net photosynthetic CO_2 fixation by the interaction of CO_2, O_2 and ribulose 1,5-diphosphate carboxylase. Plant Physiol 54:678–685
13. Keck RW, Ogren WL (1976) Differential oxygen response of photosynthesis in soybean and *Panicum milioides*. Plant Physiol 58(4):552–555. https://doi.org/10.1104/pp.58.4.552
14. Tolbert N (1971) Microbodies-peroxisomes and glyoxysomes. Annu Rev Plant Physiol 22(1):45–74
15. Von Caemmerer S (2013) Steady-state models of photosynthesis. Plant Cell Environ 36(9):1617–1630. https://doi.org/10.1111/pce.12098
16. Laisk A (2022) Prying into the green blackbox. Photosynth Res 154(2):89–112. https://doi.org/10.1007/s11120-022-00960-5
17. Ethier GJ, Livingston NJ (2004) On the need to incorporate sensitivity to CO_2 transfer conductance into the Farquhar-von Caemmerer-Berry leaf photosynthesis model. Plant Cell Environ 27:137–153

18. Tholen D, Ethier G, Genty B, Pepin S, Zhu XG (2012) Variable mesophyll conductance revisited: theoretical background and experimental implications. Plant Cell Environ 35. https://doi.org/10.1111/j.1365-3040.2012.02538.x
19. Laisk A (1977) Kinetics of photosynthesis and photorespiration in C_3 plants. Nauka, Moscow
20. Kok B (1948) A critical consideration of the quantum yield of *Chlorella*-photosynthesis. Enzymologia 13:1–56
21. Yin X, Struik PC, Romero P, Harbinson J, Evers JB, Van Der Putten PEL, Vos J (2009) Using combined measurements of gas exchange and chlorophyll fluorescence to estimate parameters of a biochemical C_3 photosynthesis model: a critical appraisal and a new integrated approach applied to leaves in a wheat (*Triticum aestivum*). Plant Cell Environ 32:448–464. https://doi.org/10.1111/j.1365-3040.2009.01934.x
22. Yin X, Sun Z, Struik PC, Gu J (2011) Evaluating a new method to estimate the rate of leaf respiration in the light by analysis of combined gas exchange and chlorophyll fluorescence measurements. J Exp Bot 62:3489–3499. https://doi.org/10.1093/jxb/err038
23. Loreto F, Velikova V, Di Marco G (2001) Respiration in the light measured by $^{12}CO_2$ emission in $^{13}CO_2$ atmosphere in maize leaves. Aust J Plant Physiol 28. https://doi.org/10.1071/pp01091
24. Pinelli P, Loreto F (2003) $^{12}CO_2$ emission from different metabolic pathways measured in illuminated and darkened C_3 and C_4 leaves at low, atmospheric and elevated CO_2 concentration. J Exp Bot 54. https://doi.org/10.1093/jxb/erg187
25. Long SP, Bernacchi CJ (2003) Gas exchange measurements, what can they tell us about the underlying limitations to photosynthesis? Procedures and sources of error. J Exp Bot 54(392):2393–2401. https://doi.org/10.1093/jxb/erg262
26. Walker BJ, Ort DR (2015) Improved method for measuring the apparent CO_2 photocompensation point resolves the impact of multiple internal conductances to CO_2 to net gas exchange. Plant Cell Environ 38:2462–2474. https://doi.org/10.1111/pce.12562
27. Walker BJ, Skabelund DC, Busch FA, Ort DR (2016) An improved approach for measuring the impact of multiple CO_2 conductances on the apparent photorespiratory CO_2 compensation point through slope-intercept regression. Plant Cell Environ 39:1198–1203. https://doi.org/10.1111/pce.12722
28. Saathoff AJ, Welles J (2021) Gas exchange measurements in the unsteady state. Plant Cell Environ 44:3509–3523. https://doi.org/10.1111/pce.14178
29. Schmiege SC, Sharkey TD, Walker B, Hammer J, Way DA (2023) Laisk measurements in the non-steady state: two tests in plants exposed to warming and variable CO_2 concentrations. Plant Physiol 193(2): 1045–1057
30. Ayub G, Smith RA, Tissue DT, Atkin OK (2011) Impacts of drought on leaf respiration in darkness and light in *Eucalyptus saligna* exposed to industrial-age atmospheric CO_2 and growth temperature. New Phytol 190: 1003–1018. https://doi.org/10.1111/j.1469-8137.2011.03673.x
31. Villar R, Held AA, Merino J (1994) Comparison of methods to estimate dark respiration in the light in leaves of two woody species. Plant Physiol 105. https://doi.org/10.1104/pp.105.1.167
32. Kirschbaum MU, Farquhar GD (1987) Investigation of the CO_2 dependence of quantum yield and respiration in *Eucalyptus pauciflora*. Plant Physiol 83:1032–1036. https://doi.org/10.1104/PP.83.4.1032
33. Sharkey TD, Preiser AL, Weraduwage SM, Gog L (2020) Source of ^{12}C in Calvin-Benson cycle intermediates and isoprene emitted from plant leaves fed with $^{13}CO_2$. Biochem J 477: 3237–3252. https://doi.org/10.46678/pb.20.1046526
34. von Caemmerer S, Farquhar GD (1981) Some relationships between the biochemistry of photosynthesis and the gas exchange of leaves. Planta 153:376–387
35. McClain AM, Sharkey TD (2020) Building a better equation for electron transport estimated from Chl fluorescence: accounting for nonphotosynthetic light absorption. New Phytol 225(2):604–608. https://doi.org/10.1111/nph.16255
36. Campbell G, Norman J (1998) An introduction to environmental biophysics, 2nd edn. Springer, New York

Chapter 13

Measurement of Photorespiratory Oxygen Exchange by Membrane Inlet Mass Spectrometry

Steven M. Driever

Abstract

Photosynthesis and metabolism in plants involve oxygen as both a product and substrate. Oxygen is taken up during photorespiration and respiration and produced through water splitting during photosynthesis. To distinguish between processes that produce or consume O_2 in leaves, isotope mass separation and detection by mass spectrometry allows measurement of evolution and uptake of O_2 as well as CO_2 uptake. This chapter describes how to calculate the rate of Rubisco oxygenation and carboxylation from in vivo gas exchange of stable isotopes of $^{16}O_2$ and $^{18}O_2$ with a closed cuvette system for leaf discs and membrane inlet mass spectrometry.

Key words Oxygen exchange, Stable isotopes, Membrane inlet mass spectrometry, Gas exchange, Oxygenation, Carboxylation

1 Introduction

Plants produce and consume oxygen in various processes and tissues, such as photosynthetic water splitting, photorespiratory oxygenation, and mitochondrial respiration. The traditional method for studying net oxygen exchange involves using a Clark-type O_2 electrode to measure changes in net O_2 exchange of leaf discs in a closed cuvette [1], along with simultaneous measurement of chlorophyll fluorescence [2]. More recent techniques include photoacoustic spectrometry, electron paramagnetic resonance oximetry [3], and electrochemical fuel cell-based differential oxygen analyzers [4, 5]. However, net O_2 exchange does not allow for differentiation between outgoing and incoming fluxes of O_2 under the same conditions, such as in the light.

To study O_2 consuming and producing processes in leaves simultaneously, stable isotope forms of oxygen with different masses, ^{16}O and ^{18}O, can be used. When an atmosphere of mainly $^{18}O_2$ is created, $^{16}O_2$ increases in the presence of light and

Berkley J. Walker (ed.), *Photorespiration: Methods and Protocols*, Methods in Molecular Biology, vol. 2792, https://doi.org/10.1007/978-1-0716-3802-6_13,

originates from water splitting due to photochemistry in the leaf, while decreases in $^{18}O_2$ result primarily from O_2-consuming processes such as photorespiration, mitochondrial respiration, and other O_2-consuming processes such as the Mehler reaction. A mass spectrometer is required to distinguish between the concentration changes of $^{16}O_2$ and $^{18}O_2$. Although quite specialized, this technique provides insight into various underlying processes, including photorespiration [6, 7], respiration [8], and oxygen-reducing processes such as the Mehler-reaction [9–12], and is particularly powerful when combined with simultaneous measurement of CO_2 gas exchange and chlorophyll fluorescence. While this technique is applicable to C_3, CAM, and C_4 plants [13–15], the focus in this chapter is solely on C_3 plants.

The Rubisco rates of oxygenation (v_o) and carboxylation (v_c) by measurement of O_2 exchange can be estimated according to Renou et al. [16]. However, in contrast to the O_2 production by photochemistry, the O_2 uptake is due to multiple processes. To estimate v_o and v_c from O_2 exchange, it has to be assumed that Rubisco oxygenase activity is the main process of leaf O_2 uptake, the mitochondrial respiration in the light is similar to that in the dark [6, 17], and that O_2 uptake by other processes is negligible [17]. In addition, the O_2 uptake by Rubisco is assumed to consume 1 mole of atmospheric O_2 per two oxygenation reactions of Rubisco [18]. With this, v_o can be determined as:

$$v_o = 2/3\left(U_o^{\text{light}} - U_o^{\text{dark}}\right) \tag{1}$$

where U_o^{light} and U_o^{dark} are the uptake of $^{18}O_2$ in the light and dark, respectively. When all electrons from the photosynthetic electron transport chain result in production of NADPH for carboxylation and oxygenation reactions, v_c can be determined as:

$$v_c = E_o - v_o \tag{2}$$

where E_o is the $^{16}O_2$ evolution in the light.

This chapter describes a procedure to determine O_2 uptake (U_o) and evolution (E_o) of leaf discs using $^{18}O_2$ and membrane inlet mass spectrometry (MIMS), to determine the rates of Rubisco oxygenation (v_o) and carboxylation (v_c). Although the focus of the described procedure is on O_2 exchange, this method can be adapted to include other isotopologues (such as $^{13}CO_2$) to study photorespiration (*see* **Note 1**).

2 Materials

2.1 Plant Material

Grow plants of French bean (*Phaseolus vulgaris*) from seed in compost-filled 1 L pots in a controlled environment cabinet with a 12-h day-length at 20 °C day- and 18 °C night temperature and

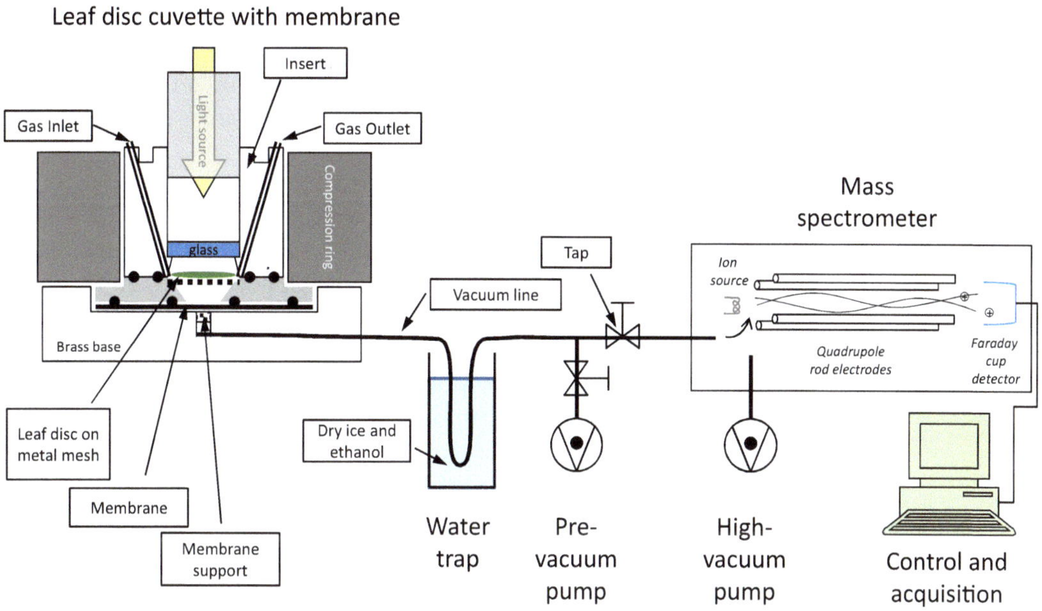

Fig. 1 A diagram of the set-up for the leaf cuvette, vacuum line, water trap, and mass spectrometer is provided. The vacuum line runs through a water trap, which contains dry ice and ethanol, in order to prevent water from entering the mass spectrometer. The tubing is initially vacuumed using the pre-vacuum pump. Once the tubing is under vacuum, the tap leading to the mass spectrometer is opened. The gases from the leaf cuvette are transferred through a membrane into the vacuum line, and then introduced into the mass spectrometer. The ionization of gases is performed by the ion source, and the mass separation is done using quadrupole rod electrodes. Ions of different masses are detected by a Faraday cup detector. The mass spectrometer is operated and data is collected using a control unit and computer system

light intensity of 300 μmol m^{-2} s^{-1} photosynthetically active photon flux density (PPFD). Maintain air humidity at an air vapor pressure deficit of ca. 0.75 kPa and ensure sufficient nutrient availability by watering regularly with Hoagland's nutrient solution [19]. Choose only fully expanded leaves for experiments described here.

2.2 Cuvette

The purpose-built, homemade measurement cuvette is designed to accommodate the measurement of O_2 exchange of a leaf disc in a small closed volume (Fig. 1), based on a design by Maxwell et al. [13] with modifications, as described by Driever and Baker [20]. In brief, the cuvette consists of a temperature-controlled brass base that supports a gas-permeable membrane, which is held in place by a steel plate with a central hole and nitrile o-rings and a large compression ring. A small porous Teflon support beneath the membrane allows gas diffusion through the membrane into the vacuum line of the mass spectrometer. The leaf disc is positioned on a metal mesh over the central hole. The cuvette is closed gas-tight by a removable steel insert with a glass window, which is screwed onto the metal plate. Actinic light is introduced by an optic fiber, which is placed in the central hole of the insert over the glass window. For injection of gases, the insert has a septum-sealed inlet and outlet, which allow rapid flushing, injection, and correction of pressure changes when introducing the desired gases into the closed cuvette.

2.3 Vacuum Line, Water Trap, and Mass Spectrometer

Gas molecules move from the cuvette to the vacuum of a mass spectrometer via a stainless-steel tube. They go through a water trap filled with dry ice and ethanol, which permits O_2 and CO_2 to pass while retaining water (*see* **Note 2** and Fig. 1). The gas molecules are ionized by a cross-beam ion source containing a tungsten filament, and a quadrupole mass spectrometer detects their subsequent masses. The mass spectrometer uses four-rod electrodes that alternate in polarity, controlled by a high-frequency generator to separate ions by mass-to-charge ratio, collected by a Faraday cup, and their signals are recorded every 3 s. During experiments, the masses m/e 28 (N_2), 32 ($^{16}O_2$), 36 ($^{18}O_2$), 40 (Ar), and 44 (CO_2) are measured. Furthermore, m/e 5.5 is measured as an indicator of signal stability, and data with excessive changes in m/e 5.5 are discarded. To minimize signal noise, the ratios of m/e 32, 36, and 44 to m/e 40 are calculated.

3 Methods

3.1 Measurement Protocol Overview

1. Select a fully expanded leaf and take a leaf disc of appropriate size (*see* Subheading 3.2 for choosing disc size).
2. Insert the leaf disc into the measurement cuvette, create an $^{18}O_2$ atmosphere (18–21%) and high CO_2 concentration (ca. 6000 μmol mol^{-1}, saturating for photosynthesis in a non-milxed, closed cuvette).
3. Start recording of changes of gases in the cuvette via a membrane inlet mass spectrometer. During an initial dark period of 10 min, determine rates of oxygen uptake in the dark (U_o^{dark}). Expose the leaf disc to a pre-defined PPFD of 500 μmol m^{-2} s^{-1} (*see* **Note 3**) and determine CO_2 assimilation rate (A), oxygen evolution (E_o), and oxygen uptake in the light (U_o^{light}) until CO_2 is depleted. Repeat this procedure with a minimum of 3 leaf discs (biological replicates).

3.2 Choosing Leaf Disc Size

To ensure optimal measurement signal over time, determine the size of leaf disc needed using pilot experiments following the procedure above but without $^{18}O_2$. Vary the leaf disc size (0.25–0.9 cm^2) and follow CO_2 uptake over time from saturating CO_2 to the CO_2 compensation point at a given PPFD. The measurement period from saturating CO_2 to compensation point should be approximately 15 min to avoid high humidity and condensation inside the cuvette. Note that the diameter of the leaf disc must always be a smaller diameter than the metal mesh of the leaf cuvette.

3.3 Membrane Installation, Calibration, Effective Cuvette Volume, and Draw Down

1. Cut the membrane to match the cuvette base size. Apply a thin layer of high vacuum grease around the outer perimeter of the brass base and ensure the central Teflon membrane support is not contaminated (*see* **Note 4**).
2. Apply the membrane to the base and smooth out any wrinkles.
3. Install the steel plate and screw on the large compression ring.
4. Check for any leaks and ensure the vacuum is stable by creating a vacuum in the vacuum line tubing that is near to that of the mass spectrometer using the pre-vacuum pump. Allow the membrane to settle onto the porous Teflon support for at least 3 h (or overnight) before starting measurements.
5. Once pre-vacuum is achieved and the membrane has settled, open the vacuum line to the mass spectrometer and allow the entire line to return to operational vacuum (according to mass spectrometer manufacturer's specification).
6. Using N_2, CO_2, Ar, $^{16}O_2$, and $^{18}O_2$, adjust the peaks for the masses of interest (according to manufacturer's instruction for the mass spectrometer used). Ensure that the sensitivities for $^{16}O_2$ (m/e 32) and $^{18}O_2$ (m/e 36) are equal to calibrate both masses to $^{16}O_2$. For handling $^{18}O_2$ gas, *see* **Note 5**.
7. To calibrate signals it is recommended to use two or more premixed gasses with CO_2 and O_2 concentrations in the range of measurement (0–2% for CO_2 and 0–21% for O_2) and pure N_2 gas for zero values. Gas mixture concentrations can vary, but should include concentrations around 400, 10,000, and 20,000 ppm for CO_2 and 2, 10, and 21% for O_2. Create a linear calibration curve for each gas concentration against its corresponding mass peak intensity signal using these known premixed gases and zero. With all calibrations, ensure a filter paper disc (mimicking a leaf disc) is present in the cuvette, to correct for volume effects of the leaf disc.
8. Calculate the effective cuvette volume by injecting a known volume of Argon (Ar) gas into the cuvette (filled with air) through the inlet port using a gas-tight syringe. The difference in the signal of the injected Ar can then be utilized to calculate cuvette volume cuvette volume (V) as:

$$V = \frac{v_{\text{inj}}}{\frac{0.0093*b}{a} - 0.0093}, \qquad (3)$$

where v_{inj} is the injected volume of Ar (mL), a is the concentration or signal of Ar in air, b is the concentration or signal of

Ar after injection, and 0.0093 is the mole fraction of Ar. This procedure is best done with a cuvette flushed with compressed air and an appropriate-sized filter paper or leaf disc (in the darkness). Mass peak intensity signals are determined when the cuvette is closed.

9. Account for draw down of gas by the system during an experiment, as a fraction of gas is drawn into the mass spectrometer vacuum. To correct for this, the draw down rate is determined for different CO_2, $^{16}O_2$, and $^{18}O_2$ concentrations during a period of ~20 min, to establish a linear relationship of draw down rate (signal or concentration min^{-1}) for each gas based upon their respective initial signal/concentration. The correction is given in the experiment described below (Subheading 4).

4 Measurement

1. Take a leaf disc from a fully expanded leaf and immediately cover the outer edges of the disc by placing it between two wet filter paper rings (*see* **Note 6**), in the cuvette on the metal mesh support.
2. Close the cuvette with the inset and purge the cuvette with N_2 gas for several minutes until $^{16}O_2$ and CO_2 are depleted. Inject $^{18}O_2$ (95% isotopic enrichment), Ar, and CO_2 to create the desired gas mixture (21% O_2 and >1% CO_2 and Ar in N_2). Ensure instantaneous pressure correction for each injection into the gas inlet by connecting the gas outlet to a tube with its end immersed in water. A second injection of CO_2 after several minutes into the measurement can be given to assure a saturating CO_2 concentration (>1%), if the CO_2 level rapidly drops below saturating level (*indicated by arrow,* Fig. 2).
3. Close the gas inlet and outlet and record changes of gases in darkness for 10 min.
4. Turn on the actinic light and continue recording the changes in CO_2, $^{18}O_2$, and $^{16}O_2$ for another 15 min. Typical exchanges of CO_2, $^{18}O_2$, and $^{16}O_2$ for French beans are shown in Fig. 2.
5. After CO_2 depletion to the compensation point (*see* Fig. 2, *indicated by* *), continue the measurement for another 5 min and stop and purge the cuvette with N_2 gas.
6. Repeat the procedure with a wet filter paper disc (of the size of the leaf disc) with gas mixtures corresponding to those used at saturating CO_2 concentration and the CO_2 compensation point, for leakage and draw down correction of gases.

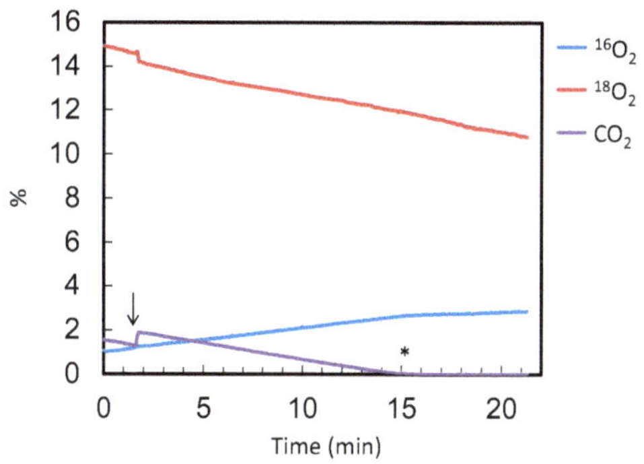

Fig. 2 Representation of expected changes of $^{16}O_2$, $^{18}O_2$, and CO_2 during gas exchange in illuminated leaf discs in a closed cuvette system using membrane inlet mass spectrometry. The data displays the transitions in exchange of a French bean leaf disc, starting from a CO_2 saturated state, which reaches the CO_2 compensation point after 14 min (* indicates this point). An arrow illustrates an additional injection of CO_2 into the cuvette. The evolution of O_2 is shown as an increase in $^{16}O_2$ and the uptake of O_2 as a decrease in $^{18}O_2$

5 Analysis

The oxygen evolution and uptake by leaf discs are calculated based on a simple model that accounts for the consumption of atmospheric oxygen and the production of oxygen from water. The production of oxygen from water is mainly observed as $^{16}O_2$ due to the low natural abundance of $^{18}O_2$ in water (0.2%). The consumption of atmospheric oxygen is a combination of three oxygen species: $^{16}O_2$, $^{16}O^{18}O$, and $^{18}O_2$. When $^{18}O_2$ is used, as in this method, the sum of the three species is primarily determined by $^{18}O_2$, allowing for a distinction between oxygen evolution producing $^{16}O_2$ and oxygen uptake consuming $^{18}O_2$. The contribution of $^{16}O^{18}O$ is considered insignificant as its natural abundance in both water and $^{18}O_2$ gas is negligible.

To apply this model in a closed cuvette system, it's necessary to consider the concentration changes of both $^{16}O_2$ and $^{18}O_2$ over time, taking into account both oxygen evolution and uptake. It's important to note that these concentration changes should be adjusted for factors such as draw down, leaks, and the effective size of the cuvette.

1. Calculation of oxygen uptake and evolution:

   Oxygen uptake (U_o) and evolution (E_o) in a closed system are calculated according to Radmer and Kok [21]. U_o describes the uptake of $^{18}O_2$ by the leaf disc over time, where the changes in $^{18}O_2$, $\partial^{18}O_2/\partial t$, are corrected for the relative abundance of $^{18}O_2$ compared to $^{16}O_2$ $(-1+{}^{16}O_2/{}^{18}O_2)$. E_o is calculated from the changes in $^{16}O_2$ over time, $\partial^{16}O_2/\partial t$, minus the changes in $^{18}O_2$, $\partial^{18}O_2/\partial t$, again corrected for their relative concentration to $^{16}O_2$:

$$U_o = -\left(1 + \frac{^{16}O_2}{^{18}O_2}\right)\frac{\partial^{18}O_2}{\partial t} \tag{4}$$

$$E_o = \frac{\partial^{16}O_2}{\partial t} - \left(\frac{^{16}O_2}{^{18}O_2}\right)\frac{\partial^{18}O_2}{\partial t} \tag{5}$$

where $^{16}O_2$ and $^{18}O_2$ are the concentrations at a given time point, corrected for draw down, leaks, and effective cuvette size. $\partial^{16}O_2/\partial t$ and $\partial^{18}O_2/\partial t$ are their first derivatives.

2. Calculation of CO_2 assimilation:

   Assimilation of CO_2 (A) in a closed system is calculated as:

$$A = \frac{\partial CO_2}{\partial t} \tag{6}$$

   where CO_2 is the concentration of CO_2 at a given time point, corrected for draw down, leaks, and effective cuvette size. $\partial CO_2/\partial t$ is the first derivative.

3. Analysis of v_o and v_c:

   Using Eqs. 4 and 5, the uptake of O_2 in the dark and light and the evolution of O_2 in the light are determined. Subsequently, v_o is calculated using Eq. 1 and v_c using Eq. 2.

6 Notes

1. To evaluate photorespiration from the uptake and evolution of CO_2, using $^{13}CO_2$, the methodology can be used as described in detail by Busch et al. [22].

2. To prepare a water trap with dry ice and ethanol, place the trap line in a dewar and fill it ¾ full with dry ice. Then carefully add ethanol (>95% pure) until all dry ice is submerged. Ethanol and dry ice will evaporate, so top up with more ethanol when dry ice starts to emerge over time (approximately after 3 h). Wear safety goggles and insulated gloves when preparing the water trap and adding the ethanol.

3. Depending on the species and growth, the appropriate, (near-) saturating PPFD during measurement can be chosen from a light response curve with an open gas exchange system. The procedure for performing a light response curve is described by Coe and Lin [23].

4. Apply silicone-based vacuum grease with a fingertip to the outside perimeter of the brass base, just enough so the membrane sticks to the base. Don't contaminate the Teflon membrane support. If there is contamination, then replace the Teflon membrane support with a new one.
5. Pure grade $^{18}O_2$ gas (>98%) is costly and therefore best used in low quantities in a small volume cuvette (as described in this chapter). To use gas sparingly and avoid dilution, a cylinder fitted with a headspace unit is used. This headspace unit consists of a four-way cross fitting, connecting the cylinder to a tube with septum, a pressure gauge, and two valves (at sides with and without septum).

   To remove air from the headspace and obtain $^{18}O_2$ gas to inject:

   - Check pressure gauge: when reading (close to) 0, there is air in the headspace. Proceed to remove the air with a vacuum pump.
   - Connect side without septum to the pre-vacuum pump.
   - Open both headspace valves.
   - Turn on pump and draw down headspace volume to vacuum.
   - Close both headspace valves tight.
   - Turn off pump.
   - Open main cylinder valve and increase pressure just above atmospheric (~1 bar).
   - Close main cylinder valve.
   - Put needle of gas-tight syringe (luer lock-type) through septum, open headspace valve at septum side, and see pressure (on gauge) drop when purging gas
   - Close the headspace valve, close syringe lock, and remove gas syringe needle from septum.

   To save $^{18}O_2$ gas from the headspace back into the main cylinder (wear safety glasses and insulated gloves):

   - Plunge cylinder into liquid nitrogen, wait until fully cooled (liquid nitrogen stops fizzing).
   - Open main cylinder valve.
   - Watch pressure drop on gauge (the leftover $^{18}O_2$ gas in the headspace shoots back into the cylinder).
   - Close main valve and let the cylinder warm up to room temperature.

6. The leaf discs are sandwiched between two moist filter paper rings that cover the outer edges of the discs to prevent drying out. Similar in diameter to the outer edge of the leaf disc, the wet filter paper rings are used instead of a single wet filter paper disc to enable free gas diffusion from both sides of the disc. The leaf disc's outer edge is held between the two moist filter paper rings, ensuring adequate hydration of the entire edge of the leaf disc during the measurement period.

References

1. Delieu T, Walker DA (1981) Polarographic measurement of photosynthetic oxygen evolution by leaf-disks. New Phytol 89(2):165–178. https://doi.org/10.1111/j.1469-8137.1981.tb07480.x
2. Delieu TJ, Walker DA (1983) Simultaneous measurement of oxygen evolution and chlorophyll fluorescence from leaf pieces. Plant Physiol 73(3):534–541. https://doi.org/10.1104/Pp.73.3.534
3. van Gorkom HJ, Gast P (1996) Measurement of photosynthetic oxygen evolution. In: Biophysical techniques in photosynthesis. Springer, pp 391–405
4. Davey PA, Hunt S, Hymus GJ, DeLucia EH, Drake BG, Karnosky DF, Long SP (2004) Respiratory oxygen uptake is not decreased by an instantaneous elevation of [CO_2], but is increased with long-term growth in the field at elevated [CO_2]. Plant Physiol 134(1): 520–527. https://doi.org/10.1104/pp.103.030569
5. Willms JR, Dowling AN, Dong ZM, Hunt S, Shelp BJ, Layzell DB (1997) The simultaneous measurement of low rates of CO_2 and O_2 exchange in biological systems. Anal Biochem 254(2):272–282. https://doi.org/10.1006/abio.1997.2416
6. Cousins AB, Pracharoenwattana I, Zhou WX, Smith SM, Badger MR (2008) Peroxisomal malate dehydrogenase is not essential for photorespiration in Arabidopsis but its absence causes an increase in the stoichiometry of photorespiratory CO_2 release. Plant Physiol 148(2):786–795. https://doi.org/10.1104/pp.108.122622
7. Cousins AB, Walker BJ, Pracharoenwattana I, Smith SM, Badger MR (2011) Peroxisomal hydroxypyruvate reductase is not essential for photorespiration in Arabidopsis but its absence causes an increase in the stoichiometry of photorespiratory CO_2 release. Photosynth Res 108(2–3):91–100. https://doi.org/10.1007/s11120-011-9651-3
8. Walker BJ, Cousins AB (2013) Influence of temperature on measurements of the CO_2 compensation point: differences between the Laisk and O_2-exchange methods. J Exp Bot 64(7):1893–1905. https://doi.org/10.1093/jxb/ert058
9. Biehler K, Haupt S, Beckmann J, Fock H, Becker TW (1997) Simultaneous CO_2- and $^{16}O_2/^{18}O_2$-gas exchange and fluorescence measurements indicate differences in light energy dissipation between the wild type and the phytochrome-deficient aurea mutant of tomato during water stress. J Exp Bot 48(312):1439–1449. https://doi.org/10.1093/jxb/48.7.1439
10. Driever SM, Baker NR (2011) The water-water cycle in leaves is not a major alternative electron sink for dissipation of excess excitation energy when CO_2 assimilation is restricted. Plant Cell Environ 34(5):837–846. https://doi.org/10.1111/j.1365-3040.2011.02288.x
11. Haupt-Herting S, Fock HP (2000) Exchange of oxygen and its role in energy dissipation during drought stress in tomato plants. Physiol Plantarum 110(4):489–495. https://doi.org/10.1111/j.1399-3054.2000.1100410.x
12. Haupt-Herting S, Fock HP (2002) Oxygen exchange in relation to carbon assimilation in water-stressed leaves during photosynthesis. Ann Bot-London 89:851–859. https://doi.org/10.1093/aob/mcf023
13. Maxwell K, Badger MR, Osmond CB (1998) A comparison of CO_2 and O_2 exchange patterns and the relationship with chlorophyll fluorescence during photosynthesis in C3 and CAM plants. Aust J Plant Physiol 25(1):45–52
14. Shirao M, Kuroki S, Kaneko K, Kinjo Y, Tsuyama M, Forster B, Takahashi S, Badger MR (2013) Gymnosperms have increased capacity for electron leakage to oxygen (Mehler and PTOX reactions) in photosynthesis compared with angiosperms. Plant Cell Physiol 54(7):1152–1163. https://doi.org/10.1093/pcp/pct066

15. Siebke K, Ghannoum O, Conroy JP, Badger MR, von Caemmerer S (2003) Photosynthetic oxygen exchange in C4 grasses: the role of oxygen as electron acceptor. Plant Cell Environ 26(12):1963–1972. https://doi.org/10.1046/j.1365-3040.2003.01112.x
16. Renou JL, Gerbaud A, Just D, Andre M (1990) Differing substomatal and chloroplastic CO2 concentrations in water-stressed wheat. Planta 182(3):415–419. https://doi.org/10.1007/BF02411393
17. Ruuska SA, Badger MR, Andrews TJ, von Caemmerer S (2000) Photosynthetic electron sinks in transgenic tobacco with reduced amounts of Rubisco: little evidence for significant Mehler reaction. J Exp Bot 51:357–368. https://doi.org/10.1093/jexbot/51.suppl_1.357
18. Badger MR (1985) Photosynthetic oxygen-exchange. Ann Rev Plant Phys 36:27–53. https://doi.org/10.1146/annurev.arplant.36.1.27
19. Arnon D, Hoagland D (1940) Crop production in artificial culture solutions and in soils with special reference to factors influencing yields and absorption of inorganic nutrients. Soil Sci 50:463–485
20. Driever SM, Baker NR (2018) Measurement of O 2 uptake and evolution in leaves in vivo using stable isotopes and membrane inlet mass spectrometry. In: Photosynthesis: methods and protocols. Humana Press, pp 141–154
21. Radmer RJ, Kok B (1976) Photoreduction of O_2 primes and replaces CO_2 assimilation. Plant Physiol 58(3):336–340. https://doi.org/10.1104/Pp.58.3.336
22. Busch FA, Deans RM, Holloway-Phillips M-M (2017) Estimation of Photorespiratory fluxes by gas exchange. In: Fernie AR, Bauwe H, Weber APM (eds) Photorespiration: methods and protocols. Springer, New York, pp 1–15. https://doi.org/10.1007/978-1-4939-7225-8_1
23. Coe RA, Lin H (2018) Light-response curves in land plants. In: Photosynthesis: methods and protocols. Springer, New York, pp 83–94

Chapter 14

Using Dynamic Gas Exchange Measurements During Oxygen Transients to Study Nonsteady-State Photorespiration

Xinyu Fu and Berkley J. Walker

Abstract

Leaf-level gas exchange is widely used to investigate the largest carbon fluxes in illuminated leaves, offering a nondestructive way to investigate the impact of photorespiration on plant carbon balance. Modern commercial gas exchange systems allow high temporal resolution measurements under changing environments, aiding the development of nonsteady-state approaches for measuring dynamic photosynthetic responses. Here, we describe a nonsteady-state technique for acquiring the dynamic response of net CO_2 assimilation to changes in photorespiratory fluxes manipulated by O_2 mole fractions. This technique allows for the screening of plant genotypes with variations in their efficiencies of photorespiration under nonsteady-state conditions.

Key words Dynamic Photosynthesis, Dynamic Photorespiration, Oxygen Transient, Gas exchange, Nonsteady-state Photorespiration

1 Introduction

Leaf-level gas exchange is commonly used to measure instantaneous rates of net CO_2 assimilation for both steady-state measurements of leaf photosynthesis under ambient conditions and for generating response curves under various conditions (e.g. light inductions). Leaf-level gas exchange has been used to estimate the rate of photorespiration in leaves and can provide valuable insights into photorespiratory metabolism [1, 2]. For example, measuring the rate of photosynthesis in a low-oxygen environment (typically 2% O_2) and comparing it to the rate of photosynthesis under normal oxygen concentrations (around 21% O_2) is commonly used to estimate the fraction of photosynthesis that is inhibited by photorespiration, although there are some deficiencies in this approach [1, 2]. Despite these deficiencies, low O_2 treatments are effective at greatly minimizing rates of photorespiration without the use of high CO_2, which impacts other aspects of plant and

Berkley J. Walker (ed.), *Photorespiration: Methods and Protocols*, Methods in Molecular Biology, vol. 2792,
https://doi.org/10.1007/978-1-0716-3802-6_14,

photosynthetic response. Alternatively, rates of rubisco oxygenation can be calculated from biochemical photosynthesis models using steady-state gas exchange measurements [3]. However, these steady-state techniques to measure the rates of photosynthesis under constant conditions do not allow the evaluation of how changes in photorespiration impact nonsteady-state photosynthesis under transient conditions.

Transient gas exchange techniques allow measuring the dynamic responses of photosynthesis to changes in environmental conditions, which can provide insights into the impact of photorespiration on nonsteady-state photosynthesis. Modern commercial gas exchange systems collect leaf gas exchange data at greater precision and higher temporal resolution, enabling measurements across transient conditions. The impact of photorespiration on photosynthesis under transient conditions is well described under some situations. For example, measuring the dynamic photosynthesis in the first few minutes after turning off the light, a method known as the post-illumination CO_2 burst, provides a qualitative estimation of photorespiration and respiration [4]. The post-illumination CO_2 burst is thought to originate mainly from the glycine metabolism that has accumulated in the leaf during photosynthesis before the light-dark transition [5]. More than a measuring technique, the post-illumination CO_2 burst illustrates that the nonsteady-state behavior of photorespiration can affect net CO_2 assimilation through photosynthesis. Specifically, the build-up and relaxation of glycine pools can directly impact the response of photosynthesis to changing environmental conditions, such as changes in light intensity and O_2 [6, 7].

When transitioning from low to high oxygen conditions, rates of net CO_2 assimilation decline at an exponential rate until reaching a lower steady state. While the transient accumulation of glycine can explain up to 20–30% of the gain in net CO_2 assimilation during the delayed photosynthetic response [8], other processes may also be involved in the transient photosynthetic response. Understanding the factors that contribute to the transient photosynthetic response is important for improving our understanding of the role of photorespiration under more physiologically relevant situations like fluctuating light. Here, we describe detailed protocols for characterizing the dynamic response of net CO_2 assimilation to changes in O_2 mole fraction, including setting up the gas exchange instrument, taking transient measurements, and modeling photosynthetic responses during oxygen transients. This quantitative approach can be used to detect otherwise subtle differences in photorespiratory-dependent photosynthetic responses during oxygen transients that would not be apparent during other transients with more physiological relevance (like light).

2 Materials

1. Infrared gas exchange analyzer (IRGA) (*see* **Note 1**).
2. CO_2 cartridges, soda lime, and desiccant for the IRGA system.
3. O_2 tank with regulator.
4. N_2 tank with regulator.
5. Tubing and a Swagelok T-fitting.
6. Two mass-flow controllers.
7. A Raspberry Pi computer with a touchscreen monitor or other way to control the proportion of gas flow through the mass flow controllers quickly.
8. C3 plants at an appropriate age for gas exchange analysis where photorespiration can be suppressed by supplying low O_2.

3 Methods

3.1 Introducing a Custom Air Supply to the Gas Exchange System

1. Use appropriate regulators, tubing, and fittings to connect the mass flow controllers to the external O_2 tank and N_2 tank (Fig. 1). The mass flow controllers can be connected to a custom Raspberry Pi computer with a touchscreen monitor (code and further details available upon request) or an alternative controller to transition between O_2 concentrations.
2. Assemble a T-fitting with one tube coming from the mass flow controllers, one tube going to the filter cover air Inlet on the back of the IRGA console (*see* **Note 2**), and the other tube (10–15 cm) venting to the ambient air to avoid overpressure into the IRGA while minimizing back diffusion of ambient atmosphere (Fig. 1).

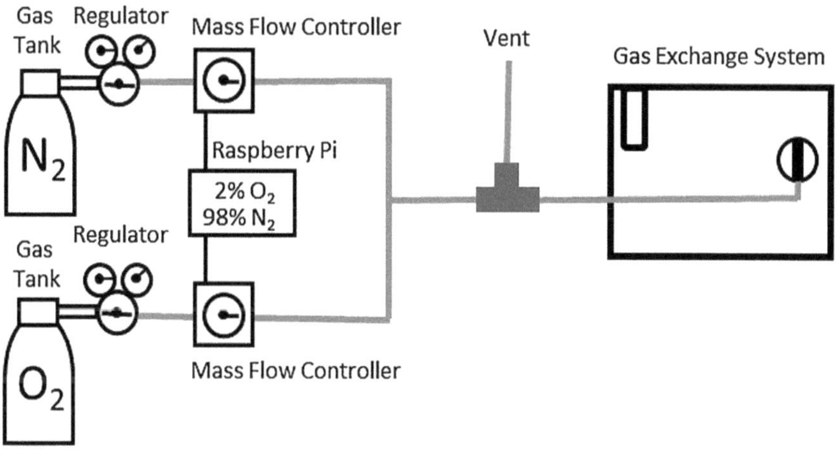

Fig. 1 Schematic showing the supply of controlled custom air to the gas exchange system. Gas flow shown in grey lines with electrical connections shown with black lines

3.2 Instrument Setup

1. Set up the gas exchange analyzer and run system warmup tests with a new CO_2 cartridge installed according to the manufacturer's instructions.
2. Close the leaf chamber and check the proper functioning of the flow control, CO_2 control, water control, temperature control, light source, and sensors.
3. Set the desired chamber condition. For example, set chamber airflow to 400 μmol/s, set chamber fan mixing speed to 10,000 rpm, control the leaf temperature that is close to the growth environments of the plants, set a light intensity that is representative of the growth environment, and control relative humidity or vapor pressure deficit in the leaf chamber to avoid condensation and stomatal closure.
4. Turn on the air supply from the mass flow controllers and adjust total airflow to be slightly higher than the flow required from the IRGA inlet. This can be ensured by checking that the vented part of the T-fitting has excess flow (*see* **Note 3**).
5. Adjust the mass flow controllers to supply low O_2 (2% O_2/98% N_2). Monitor the net CO_2 assimilation rate to ensure it increases when compared to the normal 21% O_2 atmosphere before the O_2 switch.
6. Change the system oxygen concentration value in the IRGA itself to account for the band-broadening effects of oxygen on the infra-red absorbance (and hence the measurement) of CO_2 and H_2O, this is essential for correct measurements of net assimilation under altered O_2 concentrations (*see* **Note 4**).

3.3 Measurement

1. Open a new file on the IRGA to log the data.
2. Clamp a leaf in the leaf chamber under low O_2 (2% O_2/98% N_2) with the settings outlined above. Match the reference and sample analyzer if using a dual-path IRGA.
3. The leaf should be equilibrated for 20–30 min at low O_2 until parameters such as assimilation rate, transpiration rate, and stomatal conductance become stable.
4. Manually set the valve that controls the CO_2 concentration in the IRGA to reduce oscillations in the CO_2 concentration when switching oxygen (*see* **Note 5**).
5. Set up an auto program to log data to the instrument. Set "Oxygen values" to 2, 40. Set "Log interval" to 1 s (*see* **Note 6**). Set "Step duration" to 5 min (*see* **Note 7**). The instrument will record data with oxygen constant of 2% for the first 5 min and the oxygen constant of 40% for the second 5 min.
6. Start the auto program to make the measurements. Monitor the progress of the program closely. Once the instrument has finished recording the first 5 min for low O_2 (2% O_2/98% N_2),

immediately adjust the mass flow controller to supply high O_2 (40% O_2/60% N_2) if this is already not automated at the flow rate of 2 mL/min (*see* **Note 8**).

7. Measure the leaf area when gas exchange measurements are finished if the leaf incompletely fills the measurement chamber. The leaf's interior section that was exposed within the chamber can be outlined using a permanent marker, and the outlined leaf area can be imaged and measured to obtain an accurate estimation of the leaf area.
8. Clamp the next leaf to be measured in the leaf chamber under low O_2 (2% O_2/98% N_2). Check for the settings for oxygen constant and CO_2 concentration as outlined above before new measurements.
9. When done, turn off the instrument and mass flow controller properly and close both N_2 and O_2 tanks (*see* **Note 9**).

3.4 Data Analysis

Kinetic experiments such as oxygen transitions can generate a large amount of data that is often compared qualitatively rather than quantitatively. To address this issue, we developed a pipeline to obtain quantitative information about the shape and dynamics of the oxygen transient. The pipeline may also be applied to other types of kinetic experiments to extract quantitative information about the dynamics of the system under investigation.

1. Download the data from the gas exchange analyzer to a computer and assess the quality of the collected data. Recompute data in the logged Excel Files using the measured leaf area if needed and convert the data files to CSV format. If the data is satisfactory, format the CSV file to contain a single row for the header.
2. The oxygen transient data can be described using exponential decay functions, where the net CO_2 assimilation rate decreases over time at a rate proportional to its current assimilation rate. Depending on plant species or genotypes, a one-phase or a two-phase exponential decay function can be used. One-phase exponential decay is used when the data exhibit a single exponential decay pattern, where the decay rate is constant over time. Two-phase exponential decay is used when the data exhibit a more complex pattern of decay, where the decay rate changes over time (*see* **Note 10**).

   The equation for a one-phase exponential decay function is

$$f(t) = \text{span} \cdot e^{-k \cdot t} + \text{plateau}$$

   where ***f(t)*** is the net CO_2 assimilation rate at time t, **span** is the difference between the highest and lowest possible values of net CO_2 assimilation rates, ***k*** is a constant that determines the rate of decay (also known as the rate constant), ***t*** is time, and **plateau** is the lowest possible value of net CO_2 assimilation rate in low O_2.

The equation for a two-phase exponential decay function is

$$f(t) = \text{span\_fast} \cdot e^{-k\_\text{fast} \cdot t} + \text{span\_slow} \cdot e^{-k\_\text{slow} \cdot t} + \text{plateau}$$

where ***f(t)*** is the net CO_2 assimilation rate at time *t*, **span_fast** is the difference between the initial assimilation rate and the assimilation rate when the fast decay component reaches its asymptote, **k_fast** is a constant that determines the rate of decay for the fast decay component, **span_slow** represents the difference between the assimilation rate when the fast decay component reaches its asymptote and the plateau value, **k_slow** is a constant that determines the rate of decay for the slow decay component, ***t*** is time, and **plateau** is the lowest possible value of the net CO_2 assimilation rate in low O_2.

3. To fit the data to a one or two-phase decay, download our software tool named oxygenTransientFit at https://github.com/codexf/oxygenTransientFit. The tool allows you to upload a CSV file (Fig. 2) and explore fitting exponential decay models to your data (Fig. 3). Note that there is a large spike in apparent assimilation during the first ~10 s of the measurement. This spike occurs even in an empty chamber and is very similar to the time it takes to transition entirely to

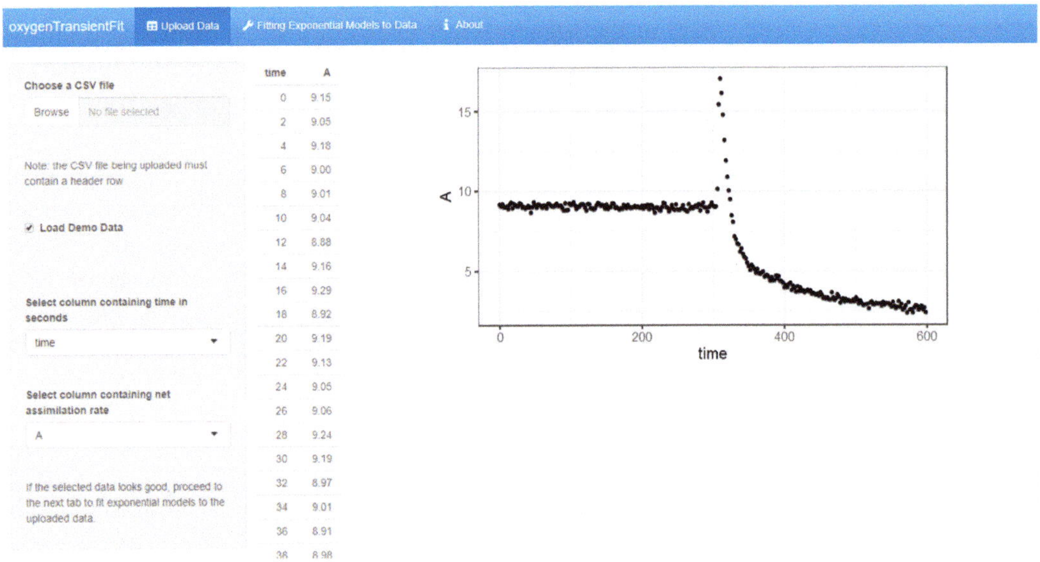

Fig. 2 The "Upload Data" tab of the oxygenTransientFit application. This tab allows users to upload their CSV data and select the columns containing the time and net CO_2 assimilation rates data. Users can also view a table of the selected data and a plot of the original data. A demo dataset is provided for users to explore fitting models

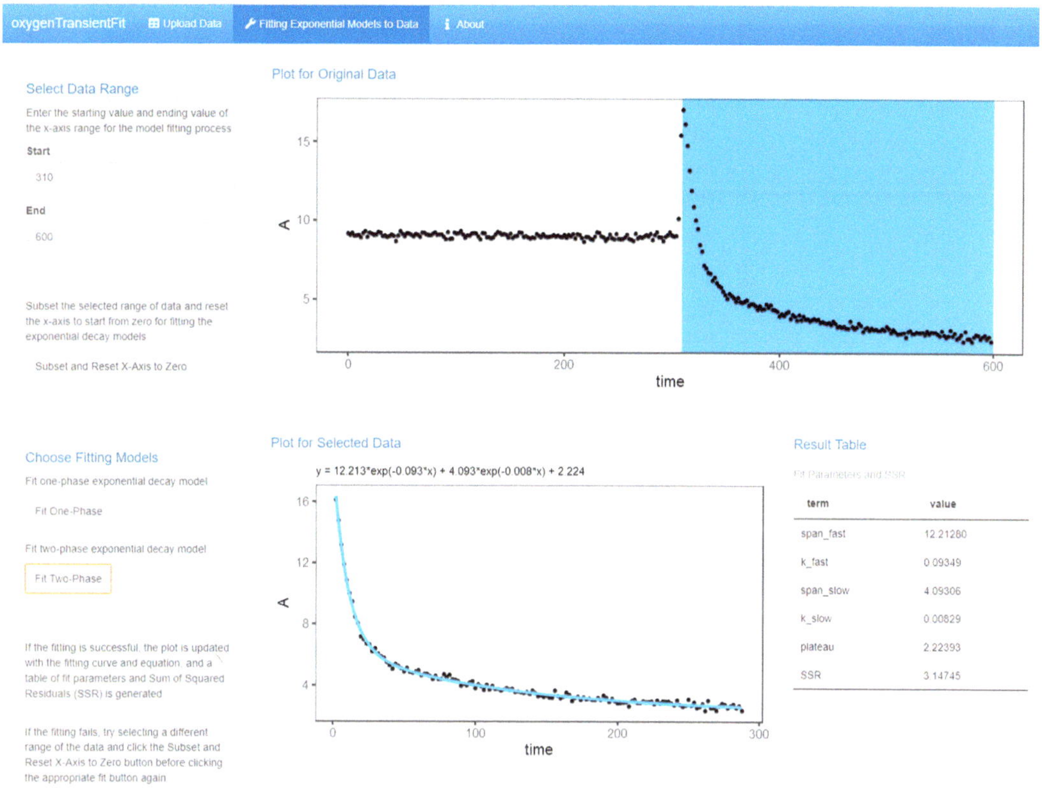

Fig. 3 The "Fitting Exponential Models to Data" tab of the oxygenTransientFit application. This tab allows users to fit one-phase or two-phase exponential decay models to their data. Users can select a range of the *x*-axis for fitting and reset the *x*-axis to a new range. The app generates a plot of the selected data with the fitted curve and a table of the fit parameters and the sum of squared residuals (SSR)

a new atmosphere using the flow rates in this experiment. This spike is most likely an artifact resulting from when the IRGA's are using the new O_2 correction for 40% O_2 despite the action chamber O_2 transitioning between 2% and 40%. Therefore, the data can be narrowed down to only include the exponential decrease phase.

4. Individual data points from each oxygen transient experiment can be fit with one of the decay models above. The rate constant is a key parameter that determines the rate at which the net CO_2 assimilation rate approaches its new steady-state value. The larger the rate constant, the faster net CO_2 assimilation rate will decay to its steady state. Comparing the rate constants of different plant genotypes can be a useful tool for identifying the underlying biological factors that contribute to delayed transient response and for developing strategies to improve photosynthetic performance in changing environments (*see* **Note 10**).

4 Notes

1. We use LI-6800 portable gas exchange system (LI-COR Biosciences, Lincoln, NE). Other models of IRGA instruments may have slight variations in their operation and output format, but the basic principle of oxygen transient measurement and data analysis still applies.
2. The LI6800 filter cover air inlet is attached to the air inlet cap via a threading mechanism and can be equipped with an M5 hosebarb. This inlet is located upstream of the pump, allowing for control of the introduced air supply by the LI-6800 pump.
3. A T-valve is essential to prevent the IRGA pump from drawing a vacuum or becoming over-pressurized. To ensure that the IRGA pump is not drawing ambient air into the inlet through the T-valve (resulting in an O_2 and N_2 mix that is different from what is desired), it is important to adjust the incoming airflow to be sufficiently high to exceed the system's demand. This is accomplished by always making sure the requested flow from the mass flow controllers is greater than the inlet flow into the IRGA. To ensure that there is excess flow, feel for a slight airstream from the inlet valve on the back of your hand or lower lip. The stream does not need to be strong, just enough to minimize potential back-diffusion of ambient air into the main airstream. Venting the excess air to the atmosphere also helps maintain the system at atmospheric pressure, which is necessary for accurate gas exchange measurements.
4. We recommend setting the oxygen constant correctly during gas exchange measurements rather than changing the constant and re-computing the data after the experiment.
5. CO_2 concentration is often controlled using a feedback loop between a control valve and the CO_2 concentration measured by the IRGA. This system works well for steady-state measurements to control CO_2, but during situations where the apparent CO_2 concentrations change, as occurs during O_2 transients, this feedback loop introduces artificial fluctuations in measured CO_2 concentrations, causing the CO_2 control valve to open or shut to compensate. This introduces unnecessary oscillations in the measured assimilation rates. To combat this, the CO_2 control valve should be set to a constant setting during the transient measurement. For example, in the LI-6800, turn the "CO_2 Injector" on under Environment >CO_2. Tap the "Injector" button and then tap the cartridge graphic. Adjust the CO_2% (open) value to achieve the ambient CO_2 concentration in the reference of sample gas analyzer.

6. If using a 1 s log interval is slowing down the instrument due to processing speed, it is sufficient to capture the oxygen transient response using a 2 s log interval instead.
7. To ensure that the leaf has reached a new steady-state assimilation rate, it is important to verify whether a 5 min data logging duration is sufficient after switching to high O_2. If the new steady-state assimilation rate has not been achieved, it may be necessary to increase the data logging duration to obtain the complete response curve.
8. If a Raspberry Pi computer is used to control the mass flow controller, it is possible to connect the Raspberry Pi with LI-6800 and configure the auto program to adjust the mass flow controller to automatically change the O_2 concentration. This will make the measurements easier and data format more consistent. Because the auto program sets the oxygen constant to 40% in the second 5 min, we do not need to set it manually.
9. To prevent running out of gas during experiments, keep a spare tank and order a replacement if the current tank is running low.
10. Based on our experience, the one-phase exponential decay model is effective for tobacco. However, for Arabidopsis, especially in certain mutants, the two-phase exponential decay model is more appropriate. The *k*_slow of some mutants with delayed transient responses is several times higher than that of the wild-type.

Acknowledgments

This work was supported by the US Department of Energy Office of Science, Basic Energy Sciences under Award DE- FG02-91ER20021 and the National Science Foundation under grant number 2030337.

References

1. Sharkey TD (1988) Estimating the rate of photorespiration in leaves. Physiol Plant 73:147–152. https://doi.org/10.1111/j.1399-3054.1988.tb09205.x
2. Busch FA (2013) Current methods for estimating the rate of photorespiration in leaves. Plant Biol 15:648–655. https://doi.org/10.1111/j.1438-8677.2012.00694.x
3. Walker BJ, Kramer DM, Fisher N, Fu X (2020) Flexibility in the energy balancing network of photosynthesis enables safe operation under changing environmental conditions. Plan Theory 9:301. https://doi.org/10.3390/plants9030301
4. Decker JP (1955) A rapid, postillumination deceleration of respiration in green leaves. Plant Physiol 30:82–84. https://doi.org/10.1104/pp.30.1.82
5. Rawsthorne S, Hylton CM (1991) The relationship between the post-illumination CO2 burst and glycine metabolism in leaves of C3 and C3-C4 intermediate species of Moricandia. Planta 186:122–126. https://doi.org/10.1007/BF00201507

6. Fu X, Walker BJ (2023) Dynamic response of photorespiration in fluctuating light environments. J Exp Bot 74:600–611. https://doi.org/10.1093/jxb/erac335
7. Fu X, Smith K, Gregory L et al (2023) Modifying photorespiration to optimize crop performance. In: Understanding and improving crop photosynthesis. Burleigh Dodds Science Publishing, Cambridge, UK, pp 203–222
8. Fu X, Gregory LM, Weise SE, Walker BJ (2023) Integrated flux and pool size analysis in plant central metabolism reveals unique roles of glycine and serine during photorespiration. Nat Plants 9:169–178. https://doi.org/10.1038/s41477-022-01294-9

Part III

Metabolic Tools

Chapter 15

Analysis of Photorespiratory Intermediates Under Transient Conditions by Mass Spectrometry

Anastasija Plett, Philipp Westhoff, and Nicole Linka

Abstract

Photorespiration is an essential process of phototropic organisms caused by the limited ability of rubisco to distinguish between CO_2 and O_2. To understand the metabolic flux through the photorespiratory pathway, we combined a mass spectrometry-based approach with a shift experiment from elevated CO_2 (3000 ppm) to ambient CO_2 (390 ppm). Here, we describe a protocol for quantifying photorespiratory intermediates, starting from plant cultivation through extraction and evaluation.

Key words Photorespiration, Mass Spectrometry, Gas chromatography

1 Introduction

The dual function of rubisco as a carboxylase as well as oxygenase leads to the production of 3-phosphoglycerate (3-PGA) as well as 2-phosphoglycolate (2-PG). While 3-PGA is further processed in the Calvin Benson Bassham cycle, 2-PG must be detoxified by the photorespiratory cycle. Changing (high-to-ambient) CO_2 transitions can induce photorespiration [1–3]. Therefore, those conditions can help to characterize plant mutants that are involved in photorespiration but lack a pronounced visible phenotype.

Gas chromatography coupled with mass spectrometry (GC-MS) is a reliable method for measuring most intermediates in the photorespiratory cycle. This analytical system is robust and versatile, and relatively easy to maintain compared to liquid chromatography systems. In addition, electron impact (EI) ionization results in target-specific molecular fragmentation patterns that allow or unambiguous identification of compounds, even with

---

Anastasija Plett and Philipp Westhoff contributed equally with all other contributors.

Berkley J. Walker (ed.), *Photorespiration: Methods and Protocols*, Methods in Molecular Biology, vol. 2792,
https://doi.org/10.1007/978-1-0716-3802-6_15,

"simple" MS systems such as single quadrupoles. In this chapter, we outline an experimental pipeline for GC-MS-based metabolic profiling focused on photorespiratory intermediates.

2 Materials

2.1 Reagents and Solutions

1. Sodium hypochlorite (12% active chlorine).
2. Concentrated HCL (*see* **Note 1**).
3. Ultrapure water (H_2O).
4. Solid plant cultivation medium: 2.2 g/L MS salts, 10 mL/L 0.5 M MES/KOH pH 5.7, 8 g/L plant Agar.
5. Liquid nitrogen (*see* **Note 2**).
6. Extraction solution **A**: methanol/chloroform (10: 4.28), MS grade (*see* **Note 3**).
7. Extraction solution **B**: ultrapure water containing 4.46 μM ribitol (CAS-Nr: 488-81-3) & N,N-dimethylphenylalanine (CAS-Nr: 17469-89-5) as internal standards (ISTD; *see* **Note 4**).
8. Methoxylamine hydrochloride (CAS-Nr: 593-56-6).
9. N-Methyl-N-(trimethylsilyl)trifluoroacetamide (MSTFA; CAS-Nr: 24589-78-4).
10. Hexane (for GC ≥98% purity).
11. Ethanol (absolute ≥99.8% purity).
12. Pyridine (for HPLC).

2.2 Equipment

1. Reaction tubes: 1.5 mL, 2 mL, safe-lock (*see* **Note 5**).
2. Petri dishes (round, 92 × 16 mm).
3. Surgical tape (breathable).
4. Plant growth chamber.
5. Aluminum foil.
6. Mortar and pestle.
7. Precision balance (±0.1 mg).
8. Vortex.
9. Centrifuge precooled to 4 °C.
10. Freeze dryer.
11. Vacuum centrifuge.
12. Thermomixer.
13. Crimp vial 1.5 mL.
14. Magnetical crimp cap with PTFE seal.
15. Deactivated glass insert (250 μL) with polymer foot.

16. GC-MS-System). The described procedure can be performed with any common GC-MS system (*see* **Note 6**).
17. GC Column: low bleed 5%-phenyl-methylpolysiloxane phase capillary column length: 30 m, inner diameter: 0.25 mm, film thickness 0.25 μm.

3 Methods

3.1 Plant Cultivation

1. Aliquot seeds into 2 mL reaction tubes and sterilize by surface sterilization with chlorine gas: place the seed rack with open tubes in a desiccator under the fume hood. Place the beaker with 100 mL sodium hypochlorite next to the seed rack. Add 3 mL concentrated HCL into the beaker and close the lid of the desiccator. Sterilize for 1.5–2 h.
2. Streak sterilized seeds onto ½ MS agar plates and seal with surgical tape.
3. Stratify seeds at 4 °C for 2 days.
4. Transfer plates to a plant growth chamber with elevated CO_2 concentrations (3000 ppm).
5. After 11 days, shift the seedlings to a plant growth chamber with ambient CO_2 concentrations (390 ppm) and incubate for another 3 days.

3.2 Harvest and Aliquoting of Plant Material

To prepare for harvesting, remove the surgical tape of all petri dishes at once to save time during harvest. Pre-fold aluminum foil into bags to harvest and store the material. Precool spatulas in liquid nitrogen.

1. Take the plant plates one by one out of the chamber and pour the liquid nitrogen directly (and as fast as possible) onto the petri dish and wait a few seconds until everything is frozen, including the plant cultivation medium. Scratch down the shoot tissue with a precooled spatula into the precooled aluminum bag and place the closed bag into liquid nitrogen (*see* **Note 7**).
2. Place the frozen material in a mortar and crush with pestle under liquid nitrogen. Transfer the fine powder into 2 mL safe-lock reaction tubes.
3. Weigh 20–25 mg of the fine powder and place it in a new 2 mL safe-lock reaction tube for analysis. Note the exact weight (*see* **Note 8**).

3.3 Extraction

The extraction was performed according to Arrivault et al. [4] with minor modifications. The harvested material should be kept frozen (in liquid nitrogen) to avoid sample degradation. Before extraction, precool all equipment (spatulas, reaction tubes) in liquid nitrogen, the extraction solution at −20 °C, and the H_2O on ice.

1. Place the 2 mL reaction tubes containing the sample upright in liquid nitrogen (*see* **Note 9**).
2. Add 350 μL of extraction solution **A** to every sample.
3. Take the samples out and vortex thoroughly.
4. Incubate at −20 °C for 1 h with occasional shaking (*see* **Note 10**).
5. Add 560 μL extraction solution **B.**
6. Vortex and incubate on ice for 10 min.
7. Spin down at 4 °C for 4 min at 20.000 × *g*.
8. Collect the (upper) aqueous phase in a new 2 mL reaction tube and store it on ice.
9. Add 560 μL H_2O to the previous tube and vortex.
10. Spin down at 4 °C for 4 min at 20.000 × *g*.
11. Collect the upper phase and combine it with the previous aqueous phase (**step 8**).
12. Fill the vial up to 1700 μL with H_2O (*see* **Note 11**).
13. Dry the sample by lyophilization overnight.
14. Reconstitute the sample in 500 μL ice-cold H_2O.
15. Transfer an aliquot of 10–50 μL into a glass inlet and add the same amount of methanol. The actual volumes needed might be partially system and experiment dependent and should be evaluated using a pooled sample (QC) (*see* **Note 12**). A proper sample amount allows to detect low abundant compounds such as glycolate but does not overload the column or the MS detector with high abundant signals such as glycine.
16. Dry the sample by vacuum centrifugation (approx. 4 h at 1.5 mbar or overnight)
17. The dried samples can be stored short-term at −20 °C and long-term (up to 6 months) at −80 °C.

3.4 Measurement and Evaluation

For the steps prior to injection, we recommend using a multifunctional dual-head autosampler for automated online derivatization (*see* **Note 6**). Such systems allow a time-optimized procedure, which is ideal for high-throughput measurements due to the nested sample processing together with an automatic liner exchange system. The liner should be changed every 15–20 samples, and after each change, blank samples (empty inlet derivatization) should be measured to evaluate background. In addition, a quality control (QC) sample should be measured at the beginning and end of each batch to verify system performance (*see* **Note 12**). To evaluate matrix background or matrix effects, we recommend running a chemical reference standard once per batch, with all compounds of interest as a concentration-defined pooled sample (standard mix). We strongly advise randomizing sample batches, especially if samples are processed manually, to avoid possible systematic bias.

1. Place the samples for 30 min in a vacuum centrifuge to remove any residual condensation water (*see* **Note 13**).
2. Freshly prepare 20 mg/mL methoxyamine hydrochloride in pure pyridine (*see* **Note 14**).
3. Add 10 μL methoxyamine hydrochloride to the dried sample and incubate while shaking at 37 °C for 90 min.
4. Add 90 μL MSTFA and incubate while shaking at 37 °C for 30 min.
5. Let the sample rest for 2 h at room temperature (*see* **Note 15**).
6. Inject 1 μL of derivatized sample with a GC-column flow of 1 mL/min helium. Between injections, let the autosampler needle perform wash injections with ethanol first and then hexane into a waste vial to clean the needle and to avoid needle damage (*see* **Notes 16** and **17**).
7. After injection, hold oven temperature constant at 70 °C for 2 min, ramp at 12.5 °C/min to 320 °C, and then hold constant for 5 min, resulting in a total run time of 27 min.
8. Set the MS parameters as follows: Ionize metabolites with an electron impact source at 70 eV Set mass range from 60 *m/z* to m/z 600 at 20 scans per sec.

3.5 Data Analysis

1. Evaluate measurements by screening chromatograms for background, ISTD intensity, and peak shape. With the setup described here, chromatograms can be reviewed with MassHunter Qualitative (Agilent Technologies, *see* **Note 18**, Fig. 1).

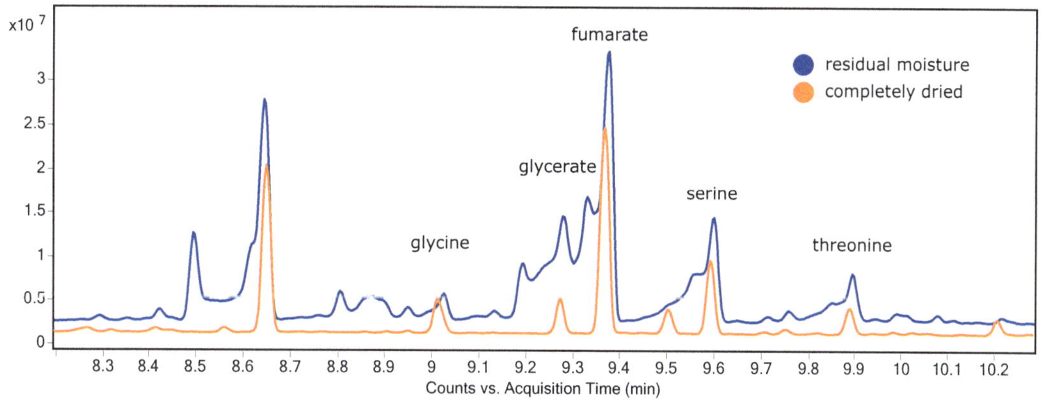

Fig. 1 Chromatogram of completely dried and partially moisture samples. Completely dried samples (orange) result in distinct peaks, which can be easily annotated. Residual moisture (blue) leads to a higher background, improper peak shapes, and double peaks, eventually leading to ambiguous interpretations

2. Identify metabolites by spectra comparison with the NIST14 Mass Spectral Library and the authentic standard.
3. To set up a data integration method in MassHunter Quantitative (Agilent Technologies), assign a quantifier ion (Q1) and a qualifier ion (Q2) for each metabolite. For the intermediates of the photorespiratory pathway, we recommend to use the following *m/z* pairs: glycolate Q1(177)/Q2 (161); glyoxylate Q1 (160)/Q2 (89); glycine Q1(174)/Q2 (248); serine Q1(204)/Q2 (218); and glycerate Q1(189)/Q2 (292) (*see* **Note 19**).
4. After integration, normalize the areas under the curve to the ISTD signal and to the fresh weight of the sample (*see* **Note 20**).

4 Notes

1. Handle under fume hood, this compound is corrosive.
2. Liquid nitrogen can freeze skin tissue, causing cold burns and or permanent eye damage. Wear gloves and safety glasses.
3. Handle under fume hood, this compound is toxic.
4. Internal standards should be carefully selected based on the target analyte and biological sample. They should reflect the physico-chemical behavior of the target metabolites during extraction and measurement but must not be present in the biological sample itself.
5. Do not autoclave, use new.
6. In this protocol, we use the multifunctional dual-head autosampler MPS for scheduled and high throughput sample preparations. However, all steps can also be performed manually in a similar manner.
7. Frozen material can be stored at −80 °C.
8. Homogenized material can be stored at −80 °C.
9. This can be easily done with a permeable reaction tube rack. The liquid nitrogen should not be in direct contact with the powder.
10. When the samples are taken out of the freezer for shaking, place them on ice.
11. The organic content in the polar phase should be less than 15% of the total volume to prevent sample loss during lyophilization. Adjust the amount of water added to the sample accordingly if the extraction volume changes.
12. We highly recommend to create a pooled sample for quality control measurements. To do this, several aliquots from

different samples reflecting all sample groups should be pooled and aliquoted in the same manner as the samples. Two quality control samples are required per GC-MS batch.

13. Derivatization reaction is very sensitive to water. Incomplete derivatization results in bad peak shapes and reduced sensitivity.
14. Weigh 20 mg dry methoxyamine hydrochloride into a 1.5 mL glass vial, add 1 mL of pure (water free) pyridine, and support complete solubilization by heating to 30 °C and shaking for 10 min.
15. The derivatization process continues even at room temperature. To ensure consistency, we recommend scheduled sample processing with an automated derivatization system. This allows each sample to be incubated for exactly the same amount of time. However, this method is impractical for manual sample handling. In such cases, we recommend working with batches of 8–10 samples and extending the resting time at room temperature to 6–8 h before measurement. This will ensure that the derivatization state is approximately the same for all samples, even if the last sample is incubated at room temperature longer than the first measured sample. However, it should be noted that incubation times may vary depending on sample composition and should be tested and optimized accordingly.
16. Here, we use a cold injection system (CIS) with an initial temperature of 50 °C. After injection, the system heats up to 250 °C with a ramp rate of 12 °C/sec. A cold injection system is recommended to standardize the injection process and lead to reproducible retention times. However, a conventional split/splitless injector can also be used. Here, we recommend an injection temperature of 230–250 °C.
17. We use baffled deactivated CIS low-volume liners for cold split injections to create good mixing of sample. For common split/splitless injectors, we recommend deactivated split liners with glass wool.
18. Agilent systems produce "*.D-files*" which are hard to read using open-source software. For that, we recommend generating ".mzXML" or ".CDF" files (e.g. with proteo wizard [5] or metalign [6]).
19. Due to the fixed electron voltage of 70 eV in common GC-MS, the fragmentation pattern is very reproducible. The ratio of Q1/Q2 can therefore be used to identify underlying masses that might interfere with accurate analysis.
20. MS response for different targets may be different for the same concentration. Therefore, the resulting relative concentrations should not be compared between different metabolites.

References

1. Timm S et al (2012) High-to-low CO 2 acclimation reveals plasticity of the photorespiratory pathway and indicates regulatory links to cellular metabolism of Arabidopsis. PLoS One 7:e42809
2. Eisenhut M et al (2017) Photorespiration is crucial for dynamic response of photosynthetic metabolism and stomatal movement to altered CO2 availability. Mol Plant 10:47–61
3. Roell M-S et al (2021) A synthetic C4 shuttle via the β -hydroxyaspartate cycle in C3 plants:1–9. https://doi.org/10.1073/pnas.2022307118
4. Arrivault S et al (2009) Use of reverse-phase liquid chromatography, linked to tandem mass spectrometry, to profile the Calvin cycle and other metabolic intermediates in Arabidopsis rosettes at different carbon dioxide concentrations. Plant J 59:826–839
5. Chambers MC et al (2012) A cross-platform toolkit for mass spectrometry and proteomics. Nat Biotechnol 30:918–920
6. Lommen A, Kools HJ (2012) MetAlign 3.0: performance enhancement by efficient use of advances in computer hardware. Metabolomics 8:719–726

Chapter 16

Combining Gas Exchange and Rapid Quenching of Leaf Tissue for Mass Spectrometry and NMR Analysis Using an External Chamber

Cyril Abadie, Julie Lalande, and Guillaume Tcherkez

Abstract

We describe here a method to study and manipulate photorespiration in intact illuminated leaves. When the CO_2/O_2 mole fraction ratio changes, instant sampling is critical, to quench leaf metabolism and thus trace rapid metabolic modification due to gaseous conditions. To do so, we combine $^{13}CO_2$ labeling and gas exchange, using a large custom leaf chamber to facilitate fast sampling by direct liquid nitrogen spraying. Moreover, the use of a high chamber surface area (about 130 cm^2) allows one to sample a large amount of leaf material to carry out ^{13}C-nuclear magnetic resonance (NMR) analysis and complementary analyses, such as isotopic analyses by high-resolution mass spectrometry (by both GC and LC-MS). ^{13}C-NMR gives access to absolute ^{13}C amounts at the specific carbon atom position in the labeled molecules and thereby provides an estimate of ^{13}C-flux of photorespiratory intermediates. Since NMR analysis is not very sensitive and can miss minor metabolites, GC or LC-MS analyses are useful to monitor metabolites at low concentrations. Furthermore, ^{13}C-NMR and high-resolution LC-MS allow to estimate isotopologue distribution in response to $^{13}CO_2$ labeling while modifying photorespiration activity.

Key words Photorespiration, ^{13}C labeling, Custom leaf chamber, Nuclear Magnetic Resonance, High-resolution mass spectrometry

1 Introduction

Common gas exchange measurements do not allow direct determination of photorespiration activity in illuminated leaves. In fact, the oxygenation rate can only be inferred using assumptions that may or may not be appropriate [1]. Net CO_2 assimilation rate (A) reflects CO_2 fixation via gross photosynthesis (carboxylation) but is also impacted by CO_2 release from photorespiration (Φ) and day respiration (R_d) associated CO_2 reassimilation [2]. In the past, some gas exchange protocols were proposed, like monitoring the post-illumination burst of CO_2 during or just after light to dark transition [3, 4]. Unfortunately, this method is qualitative and does

Berkley J. Walker (ed.), *Photorespiration: Methods and Protocols*, Methods in Molecular Biology, vol. 2792,
https://doi.org/10.1007/978-1-0716-3802-6_16,

not provide an estimate of the photorespiratory rate because CO_2 assimilation may still occur and not be completely inhibited and decarboxylated products may not only originate from photorespiration [5, 6]. Furthermore, CO_2 is released by day respiration and the transition to darkness enhances catabolic activity [7]. The CO_2 burst thus does not only reflect photorespiration and accordingly, ^{18}O labeling experiments confirmed the importance of CO_2 emissions by respiration [8, 9]. An estimate of photorespiration can also be obtained by coupling gas exchange with fluorescence measurements [10], but this approach requires assumptions concerning alternative electron sinks like nitrogen assimilation and therefore can be associated with large errors when photorespiration is low [11]. A more accurate technique to estimate photorespiration by gas exchange can be achieved by coupling an Infra-Red Gas Analyser (IRGA) system to a membrane inlet mass spectrometer and replacing $^{12}CO_2$ with $^{13}CO_2$ during net CO_2 assimilation measurements [2]. In this situation, the ^{12}C-CO_2 signal measured during a few minutes reflects decarboxylation of unlabeled photorespiration substrates. However, this approach is challenging to implement on a routine basis, requires assumptions about day respiration activity, and requires fast measurements, i.e. before photorespiratory pools become ^{13}C labeled. Alternatively, ^{18}O labeling of leaf H_2O or atmospheric O_2 can be another method to measure directly the oxygen fixation rate (V_o) [12, 13]. However, ^{18}O labeling is challenging: first, ^{18}O labeled molecules are expensive; second, ^{18}O analyses are difficult to implement and need direct monitoring with membrane inlet mass spectrometry; and third, due to isotopic exchange between water and hydroxyl and carboxyl groups, this method cannot provide reliable results for photorespiratory labeled intermediates.

Here, we outline a direct isotope-tracing method to study photorespiratory activity based on ^{13}C labeling (with $^{13}CO_2$) to follow photorespiratory intermediates and calculate metabolic fluxes [11, 14–19]. An advantage of this technique is that it does not require assumptions about fluxes that are not measured (like day respiration). Labeling experiments are assisted by a gas exchange measurement system to easily manipulate photorespiration activity by modifying the O_2/CO_2 ratio [20]. When gaseous conditions are modified in the leaf cuvette (labeling chamber), sampling can be very challenging. Therefore, using a big custom chamber separate from the gas analyzer facilitates instant sampling by liquid nitrogen spraying to quench metabolism [21]. Furthermore, a big labeling chamber allows one to get samples with sufficient leaf biomass for ^{13}C-nuclear magnetic resonance (NMR) analyses, which provide direct quantification of ^{13}C content in C-atom positions of the labeled metabolites. This isotopic method gives access to the rate of serine, glycerate, and glycine accumulation, which are photorespiratory intermediates present at relatively high concentration in leaves. Nevertheless, NMR analyses are not sensitive enough to detect all ^{13}C labeled molecules and sometimes, overlapping NMR signals prevent accurate integration [17]. Therefore, high-resolution mass spectrometry (GC and LC-MS) analysis can be undertaken to provide complementary pieces of information. For example, LC-MS analysis gives access to metabolite identification and resolution of isotopologues (^{13}C, ^{15}N, ^{33}S) and thus measures directly the relative content of ^{13}C carbon atoms assimilated during labeling and allocated to metabolites [14, 15]. Altogether, the results can be combined to calculate carbon fluxes associated with photorespiration rate and photorespiratory intermediate build-up. Leaf samples generated by the ^{13}C-based isotopic labeling method are extracted using simple perchloric extraction method for NMR [22] or classical protocols for metabolomics [23], and extracts can be analyzed using routine ^{13}C-NMR and GC and LC-MS. Here, we provide practical and methodological information on (i) the leaf chamber and how to perform gas exchange coupled to $^{13}CO_2$ labeling, (ii) sampling, (iii) NMR analyses, and (iv) mass spectrometry analyses. Further information on results, calculations, isotopic pattern resolution, etc., can be found in [11, 14–19].

2 Materials

2.1 Design of the Custom Leaf Chamber

In the following chapter, we describe how to design a custom leaf chamber and how to set up the light, temperature, and humidity inside.

1. The overall 3D design of the chamber is realized with the software Fusion 360 (Autodesk) prior to production in aluminum (*see* **Note 1**) by Computer Numerical Control milling.
2. The chamber is comprised three parts (Fig. 1): (i) the main chamber, an empty frame that welcomes the leaf (up to 130 cm^2) with a gas outlet on the side. The gas inlet is placed at the back of the chamber, where a small fan can be installed; and (ii) two aluminum frames with the same main chamber dimensions with soft and transparent plastic walls to tightly close the main chamber (*see* **Note 2**).
3. Light is provided by an LED panel (e.g. the RGBW-L084 Led panel, Heinz Walz GmbH, Effeltrich, Germany) to independently adjust light colors and to reach a maximum of 2000 $\mu mol/m^2/s$ Photosynthetically Activate Radiation (PAR).
4. Temperature is regulated by a Peltier cooling surface-to-air module placed at the rear of the custom chamber. The Peltier

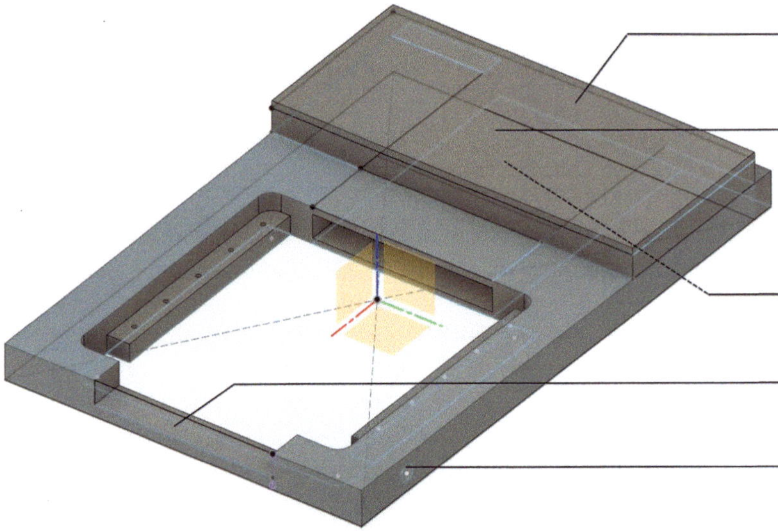

Fig. 1 3D design of the main part of the custom chamber (Fusion 360, Autodesk). The gas (incoming air) with precise O_2, CO_2, and water vapor mole fraction from the gas exchange system (Li-6800, Licor) arrives at an opening (chamber gas inlet) at the rear of the chamber. Gas temperature of the chamber is regulated by a surface-to-air Peltier module onto the chamber at the rear. A small fan inside the chamber homogenizes the air surrounding the leaf. Finally, outcoming air goes out of the custom chamber to be analyzed by the IRGA, via an opening (chamber gas outlet) on the side

module is connected to a temperature controller and an independent temperature probe fixed on the chamber (*see* **Note 3**).

5. A small flat centrifuge fan (12 V) is placed inside the cavity present at the back to homogenize the gas (air) in the main leaf chamber.
6. A homemade thermocouple is placed inside the chamber to monitor leaf temperature.
7. A rack made of aluminum extrusion bars can be used to set up and support the overall system.

2.2 Labeling Setup

1. Gas exchange system (e.g., Licor Li-6800, Li-Cor Biosciences, Lincoln, NE, USA) or Li-6400XT with dew point generator.
2. Licor-6800 Custom Chamber Adapter (6800-19).
3. Silica gel orange 2–5 mm.
4. Soda lime pellets with indicator.
5. Dreschel gas washing bottle (*see* **Note 4**).
6. CO_2 tank (10 kg CO_2 100%) (*see* **Note 5**).
7. 40 L cylinder of 99% ^{13}C-CO_2.
8. 100% O_2 tank with an O_2 regulator.
9. 100% N_2 tank with its regulator.

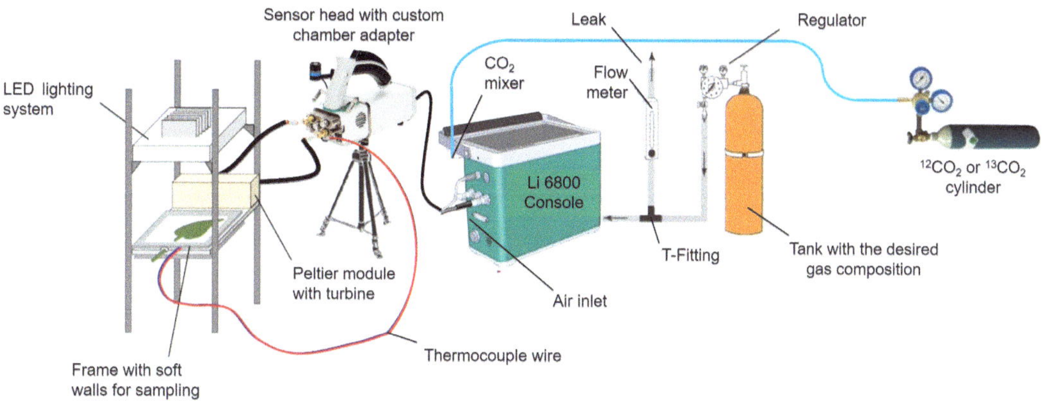

Fig. 2 Experimental setup to perform isotopic labeling using an external chamber. The system includes a portable gas exchange system allowing the control of CO_2 mole fraction by connecting a $^{12}CO_2$ or a $^{13}CO_2$ cylinder to its mixer. To modify O_2 concentration inside the chamber, a tank with the desired O_2 composition can be connected to the air inlet by using plastic tubing and a flow meter. The T-type connector between the flow meter and the air inlet allows a leak to protect the gas exchange system pump. (Redrawn from [19])

10. 2% O_2 tank.
11. Two identical two-stage regulators for CO_2 (inlet pressure range 0–300 bar and delivery pressure range 0–40 bar).
12. A gas flow meter (150 Lair/hour).
13. A diaphragm pump with adjustable flow.
14. Plastic tubing for gas connections between gas tank, flow meter, and Li-6800 air inlet (Fig. 2).
15. Licor tubing (8150-250 Bev-A-Line 1/4″, Li-Cor Biosciences, Lincoln, NE, USA) to connect the chamber gas inlet and outlet to the Li-6800 IRGAs.
16. Terostat® to ensure the tightness of the junctions (Teroson-Terostat VII, Henkel, Germany) (*see* **Note 6**).

2.3 Leaf Metabolites Extraction

1. Large mortar and pestle (400 mL).
2. 70% and 2% Perchloric acid in ultrapure water.
3. 0.25 M Maleic acid in ultrapure water.
4. 5 M Potassium hydroxide (KOH) in ultrapure water.

2.4 NMR Analysis

1. NMR spectrometer, Advance 700 MHz (Bruker BioSpin).
2. NMR Tubes 5 mm (Z107373, Bruker Biospin, Wissembourg, France).
3. Deuterium oxide (D_2O).
4. Deuterium oxide with 3-(trimethylsilyl)-1-propane-sulfonic acid (TMSPS) (*see* **Note 7**).

5. Ethylenediaminetetraacetic acid (EDTA), 75 mM in ultrapure water (*see* **Note 8**).
6. 10, 25, 50, and 100 mM Threonine solutions in ultrapure water.

2.5 GC-MS Derivation

1. GC-MS Q Exactive (Thermo Fisher Scientific).
2. TG-5 SILMS column (30 m × 0.25 mm × 0.25 μm; Thermo Scientific).
3. 20 mg/mL Methoxyamine hydrochloride for GC derivatization in pyridine (*see* **Note 9**).
4. 99.8% Pyridine anhydrous (*see* **Note 9**).
5. N-Methyl-N-trimethylsilyltrifluoroacetamide (MSTFA) (*see* **Note 9**).
6. Hexane LC-MS grade.
7. Alkane mix (at 10 μg/mL), 14 alkanes from C_9 to C_{36} in hexane (Decane C_9, Docosane C_{10}, Dodecane C_{12}, Dotriacontane C_{14}, Eicosane C_{16}, Hexacosane C_{18}, Hexadecane C_{20}, Hexatriacontane C_{22}, Nonane C_{24}, Octacosane C_{26}, Octadecane C_{28}, Tetracosane C_{30}, Tetradecane C_{32}, Tetratriacontane C_{34}, and Triacontane C_{36}) (*see* **Note 10**).
8. Perfluorotributylamine (PFTBA, FC43 calibration compound).
9. Helium gas purity 5.0 (Messer).
10. Nitrogen gas purity 4.5 (Messer).

2.6 LC-MS Analysis

1. LC-MS Orbitrap Q Exactive plus (Thermo Scientific).
2. UHPLC Ultimate 3000 LC system (Dionex-Thermo Scientific, Hemel Hempstead, UK).
3. ZIC®-HILIC column (3.5 μm, 200 Å, 150 × 2.1 mm; Merck SeQuant®).
4. ZIC®-HILIC column guard (20 × 2.1 mm; Merck SeQuant®).
5. Water LC-MS grade.
6. Acetonitrile LC-MS grade.
7. Ammonium acetate eluent additive for LC-MS (*see* **Note 11**).
8. Trifluoromethyl phenylalanine (TFMP) 5 mM in water LC-MS grade.
9. LTQ-ESI positive ion calibration solution (Pierce®, Thermo Fisher Scientific).

3 Methods

3.1 ^{13}C-CO_2 Labeling Experiment with Gas Exchange Measurements

This ^{13}C labeling protocol is carried out using a large leaf custom chamber coupled with a gas exchange measurements system allowing manipulation of leaf photorespiration via changes in gaseous conditions (with different gas tanks). This protocol allows ^{13}C-CO_2 labeling on detached leaf or directly in planta (*see* **Notes 12** and **13**).

1. Change CO_2 (soda lime) and H_2O (silica gel) in the IRGA system console daily. Silica gel can be regenerated at 120 °C for 1 h.
2. Set the CO_2 tank regulator to 15 bar.
3. Launch the Li-6800 warmup program.
4. Open a new file and start an autoprogram to log gas exchange measurements every 5 min with an additional averaging time of 3 s per log.
5. Set the following environmental conditions: sampCO_2 concentration, 400 ppm; Vapor Pressure Deficit (VPD), 0.8–1.1 KPa.
6. Match the IRGAs.
7. If required, cut the blade of the leaf to fit inside the custom chamber.
8. Measure the leaf area and note the value in the Li-6800 software to recalculate CO_2 net assimilation rate with actual leaf surface area.
9. Place the thermocouple linked to the Li-6800 inside the leaf chamber in touch with the leaf.
10. Turn the Peltier module and the fan chamber on.
11. Tighten the two frames with plastic soft walls to the main chamber with Terostat®.
12. Connect both chamber gas inlet and outlet to the Li-6800 custom chamber adapter.
13. Control pressure of the soft plastic walls to check any leaks in the chamber.
14. Cover the closed leaf chamber with aluminum foil to measure dark respiration after 10 min of gas stabilization.
15. Remove the aluminum foil and turn the light to non-limiting light intensity (*see* **Note 14**) with 10% blue light.
16. Wait 30–45 min to reach steady photosynthesis rate (*see* **Note 15**).
17. Once gas exchange rates are stabilized at the atmospheric CO_2 concentration (in this case, 400 ppm), change gas conditions to modify leaf photorespiration activity inside the chamber

Table 1
Origin of the gas used to manipulate photorespiration activity during the labeling experiment. O_2/CO_2 conditions are given in %/μmol/mol

Gas used by Li-6800 (air inlet)	O_2/CO_2 (%/ppm)	Photorespiration activity
100% O_2 Tank	100/400	Highly stimulated
Ambiant air	21/140	Stimulated
Ambiant air	21/400	Normal
Ambiant air	21/800	Decreased
2% O_2 in N_2 Tank	2/400	Highly decreased
100% N_2 Tank	0/400	Inhibited

O_2 is indicated in % of background gas (N_2)

either by modifying CO_2 concentration with the Li-6800 or O_2 concentration (*see* **Note 16**) by using a background gas with the desired O_2 composition from a tank (*see* **Note 17**). An example of conditions that can be reached and how they modify photorespiration rate is presented in Table 1.

18. Wait 10–30 min to let photosynthesis adapt to the new conditions. Adapt CO_2 reference concentration to reach the CO_2 mole fraction desired inside the chamber, CO_2 Sample.
19. As IRGAs are insensitive to $^{13}CO_2$, switching to this gas while controlling for CO_2 concentration will result in large changes to the delivered CO_2 quantity. Therefore, when unlabeled CO_2 conditions are in the steady state, note the CO_2 mixer aperture value of the Li-6800 in % (mV for Li-6400 XT) and the pressure from the CO_2 tank.
20. Start the ^{13}C-CO_2 labeling experiment by switching the CO_2 source using the ^{13}C-CO_2 cylinder. Open the regulator at the same pressure, without changing the setting of the Li-6800 CO_2 mixer (maintaining the same %) (*see* **Note 18**). When using a detached leaf, a labeling solution can be fed into the petiole. The labeled molecules can then reach the blade via the transpiration flow (*see* **Note 19**).
21. During labeling, regularly control the pressure on the ^{13}C-CO_2 cylinder regulator, as well as leaf temperature and humidity in the chamber to avoid water vapor condensation in the custom chamber or in tubing connected to the IRGA (*see* **Note 20**).
22. At the end of the labeling experiment (1–4 h), quench leaf metabolism by spraying liquid nitrogen inside the chamber through an opening cut in the plastic soft wall of the upper leaf chamber frame. Frozen leaf material can be easily sampled by cutting the soft plastic wall of the lower frame and the

petiole to drop down the labeled leaf in liquid nitrogen bath beneath the chamber.

23. For every $^{13}CO_2$ labeling experiment, repeat the same protocol with the same gas conditions with unlabeled CO_2 to quantify the sample ^{13}C natural abundance (*see* **Note 21**).

3.2 Perchloric Extraction

Leaf metabolite extraction for ^{13}C-NMR analysis does not require lengthy or complicated preparation and allows one to form aliquot for GC and LC mass spectrometry.

1. Grind frozen leaves in a mortar with liquid nitrogen.
2. Add 900 μL 70%perchloric acid.
3. Add 125 μmol of maleate as internal standard.
4. Put the frozen powder in a 50 mL conical vial.
5. Add 10 mL ultrapure water in the tube and vortex.
6. Centrifugation at 15,000 g for 15 min at 4 °C.
7. Collect the supernatant.
8. Resuspend the pellet in 3 mL 2% perchloric acid.
9. Centrifugation at 15,000 g for 15 min at 4 °C.
10. Combine both supernatants.
11. Adjust pH at 5.2 with 5 M KOH.
12. Freeze the solution in liquid nitrogen and store it at −80 °C.
13. Freeze-dry the perchloric extract solution.

3.3 ^{13}C-NMR Analyses

Both ^{13}C-labeled and unlabeled samples must be analyzed in the same conditions.

1. Resuspend dry sample with 1 mL ultrapure water with 15 mM EDTA.
2. Centrifugation at 7,000 g for 15 min at 4 °C.
3. Neutralise supernatant pH with 5 M KOH.
4. Sub-sample 20 μL for GC and LC-MS analysis.
5. Add 50 μL D_2O to 550 μL of the perchloric extract (to lock the sample for NMR) and transfer into a 5 mm diameter NMR tube.
6. ^{13}C-NMR analyses are performed with the following settings: 298 K (25 °C) without tube spinning; proton-decoupled (Waltz16) carbon pulse program (zgig) with 90° pulses for ^{13}C of 10 μs at 50 W, acquisition time, 0.9 s; FID size, 65 k; relaxation delay (D1), 1.2 s and 20,000 scans (about 12 h analysis/sample).
7. Recover the solution and keep it at −80 °C (*see* **Note 22**).
8. Integrate ^{13}C signals and normalize them to maleate.

9. To control NMR quantitatively, 10, 25, 50, and 100 mM unlabeled threonine solutions (in ultrapure water 600 μL with 50 μL D_2O or TMSPS) are analyzed on a regular basis.

3.4 GC-MS Analyses

Perchloric extracts must be derivatized to be analyzed by GC-MS.

1. Pour 2 μL of the perchloric extract into a vial with insert and spin-dry at 39 °C.
2. Sample derivatization: (i) Add 20 μL of 20 mg/mL methoxyamine in pyridine in the vial and place the vial under continuous agitation for 90 min at 37 °C; (ii) add 30 μL of N-methyl-N-(trimethylsilyl)-trifluoroacetamide (MSTFA) and place the vial under continuous agitation for 30 min at 37 °C; and (iii) before injection, add 5 μL of 10 μg/mL alkane mix in each sample to compute the retention index.
3. Inject 1 μL in splitless mode at 230 °C (injector temperature) (*see* **Note 23**). Helium is used as a gas carrier with a constant flow of 1 mL/ min.
4. Set the temperature gradient of the GC oven as follows: 1 min at initial temperature of 70 °C, then temperature rises to 325 °C at 15 °C/min to finally stay at 325 °C for 4 min.
5. Operate mass spectroscopy analyses in positive polarity in full MS scan mode with the following source settings: mass scan range, 50–750 m/z; resolution, 60,000; AGC target, 1E6; MS transfer Line, 300 °C; filament delay, 4.12 min; ionization by electron impact, 70 eV; and ion source temperature, 250 °C.
6. Control mass performances every 24 h with an EI calibration gas which contains perfluorotributylamine (PFTBA).
7. Analytes are automatically identified and integrated using TraceFinder® (Thermo Scientific).
8. Normalize integrated signals to maleate.

3.5 Targeted Analysis by Exact Mass LC-MS

1. Dilute perchloric extract 10 times with water/acetonitrile (v/v, 50:50) with trifluoromethyl phenylalanine (TFMP, 10 μM) as an internal standard for mass precision.
2. Add to the batch analysis standard samples with a mix of authentic standards (concentration adjusted to have 1 to 100 pmol per injection) of metabolites of interest (including TFMP) to carry out calibration curves and thus to control quantification conditions.
3. Set the sample tray temperature at 4 °C.
4. Inject 1 μL of the diluted extract.
5. Set the elution constant flow rate at 0.3 mL/min with a binary gradient as follows: mobile phase A is acetonitrile/water (25: 75, v/v) with 5 mM ammonium acetate and mobile phase B is

acetonitrile/water (95:5, v/v) with 5 mM ammonium acetate (*see* **Note 11**).

6. Set the solvent gradient: 72% B at 0 min, down to 36% B at 13 min, maintained for 2 min, and then back to 72% B at 16 min (total run time of 21 min).
7. MS analysis is carried out with a HESI-II probe operated in positive polarity using the full MS scan mode (*see* **Note 24**) with the following settings: source voltage, 3500 V; resolution, 70,000; AGC target, 1×10^6; mass scan range, 60–600 m/z; sheath gas, 40; auxiliary gas, 10 and sweep gas, 1.5; probe temperature, 300 °C; capillary temperature, 250 °C; S-lens RF level, 50.
8. Immediately before each analysis batch, perform mass calibration with the LTQ-ESI positive ion calibration solution.
9. Use the Xcalibur software (Thermo Scientific) to perform targeted analysis of the raw data (*see* **Note 25**).

4 Notes

1. The chamber and frames are made of aluminum to resist water and liquid nitrogen and improve temperature control.
2. The soft and transparent plastic walls (transparent polypropylene book cover film) of the two aluminum frames let the light pass through and allow instant sampling by liquid nitrogen spraying.
3. Alternatively, a water bath system can be used to control the custom leaf chamber temperature.
4. A Dreschel bottle (empty or with chemical scrubs inside like soda lime or silica gel) can be used as an interface to easily connect the Li-6800 gas inlet to the flow meter and therefore the gas tank.
5. Avoid small (8–12 gr) CO_2 cartridges for the unlabeled conditions to determine ^{13}C natural abundance because they can have an unreliable or variable ^{13}C isotope composition.
6. Terostat® application is critical to seal airtight the overall custom chamber.
7. TMSPS is useful to set the chemical shift for ^{13}C NMR analysis and improves identification of metabolites C-atom positions. The carbon atoms in TMSPS have indeed a chemical shift considered to be 0 ppm.
8. Adding EDTA in samples for ^{13}C-NMR analysis is not mandatory, but greatly improves signals resolution of –COOH groups.

9. Derivation reactions must be carried out in absence of water. Reagents must be anhydrous, therefore we recommend to purchase them in small volumes or amounts to use them quickly and avoid any water contamination.
10. In addition to the alkane mix, a Grob test can be regularly carried out by GC-MS to control column performances and check for any variations in retention times.
11. Adding ammonium acetate to elution solvents improves the separation of metabolites by liquid chromatography.
12. The overall gas exchange protocol can be easily carried out on sunflower (*Helianthus annuus*) leaves, adjustments might be needed for other species.
13. To properly detach a leaf from the plant, we recommend using a new razor blade and to recut the petiole under water prior to placing it in a 15 mL plastic test tube with water to avoid embolism in the xylem and maintain the transpiration flow.
14. Non-limiting light intensity can be determined by an initial measurement of net assimilation light-response curves with Li-6800 and observing when the response saturates. For *Helianthus annuus,* 800–1000 $\mu mol_{(photon)}/m^2/s$ is typically non-limiting.
15. As the leaf area in the custom chamber is larger (around 100 cm^2) compared to the usual leaf cuvette used with the Li-6800 (2 cm^2), the raw amount of CO_2 assimilated is greater and care must be given to adjust precisely the CO_2 concentration inside the leaf chamber (CO_2 sample). That is, CO_2 reference concentration must be increased to correct for this effect to set the correct sample CO_2 concentration.
16. Adjust the O_2 setting of the licor to properly account for the IRGA band-broadening under different O_2 concentrations.
17. Gas flow from the tank with the desired O_2 composition is controlled by its regulator, which is connected to a flowmeter with plastic tubing. Tubing from the flowmeter is then connected to the air inlet of the Li-6800. When connecting the Li-6800 gas inlet to a gas tank, allow a leak in the connections by using a T-type connector to protect the Li-6800 pump from pressurization or from pulling a vacuum if the tank empties during the experiment.
18. During the labeling experiment, CO_2 net assimilation values are not valid because IRGAs are insensitive to ^{13}C-CO_2. If H_2O inside the chamber is well set, transpiration rate and stomata conduction can be followed to control stability of gas exchange inside the leaf chamber during labeling time.
19. To use a labeling solution, remove water from the tube in which the cut petiole is placed (with a 50 mL syringe) and

replace it with the labeling solution. During the labeling time, the leaf will uptake the labeled solution via the transpiration flow; therefore, its level must be checked regularly.

20. If humidity inside the chamber is too high or if the temperature to close to the dew point, an external H_2O scrub coupled with a diaphragm pump (external drying loop) can be used to regulate H_2O inside the chamber and avoid any risk of letting liquid water inside the IRGA. However, with this solution, parameters calculated from water measurements (transpiration rate and stomata conductance) by the IRGA are no longer correct.
21. Natural ^{13}C isotope composition in plant tissues is not negligible ($\delta^{13}C \approx -28‰$ for C3 plants, i.e., about 1.09% ^{13}C). All ^{13}C labeling signals must therefore be corrected by the ^{13}C signals obtained with the control experiment that reflects ^{13}C natural abundance.
22. NMR analysis is not destructive and samples can be stored at −80 °C before later use for complementary investigations.
23. If the GC-MS analysis shows saturated signals, the sample can be re-injected in the split mode to dilute with helium flow.
24. In addition to a full MS scan, LC-MS analysis can also be conducted in the All Ion Fragmentation (AIF) mode to identify positional isotope enrichment in the fragments of interest. Note that the fragment size depends on the higher energy collision-induced dissociation (HCD) parameter.
25. Data from LC-MS can also be extracted with the software Mzmine 2 to undertake non-targeted ^{13}C metabolite analysis [16].

Acknowledgments

Authors acknowledge the support of the Région Pays de la Loire and Angers Loire Métropole via the grant Isoseed, awarded to G. T.

References

1. Busch FA (2013) Current methods for estimating the rate of photorespiration in leaves. Plant Biol 15:648–655
2. Busch FA, Sage TL, Cousins AB, Sage RF (2013) C3 plants enhance rates of photosynthesis by reassimilating photorespired and respired CO_2. Plant Cell Environ 36:200–212
3. Decker JP (1955) A rapid, postillumination deceleration of respiration in green leaves. Plant Physiol 30:82–84
4. Kebeish R, Niessen M, Thiruveedhi K, Bari R, Hirsch H-J, Rosenkranz R, Stäbler N, Schönfeld B, Kreuzaler F, Peterhänsel C (2007) Chloroplastic photorespiratory bypass increases photosynthesis and biomass production in Arabidopsis thaliana. Nat Biotechnol 25:593–599
5. Laisk A, Kiirats O, Oja V (1984) Assimilatory power (Postillumination CO_2 uptake) in leaves: measurement, environmental

dependencies, and kinetic properties. Plant Physiol 76:723–729

6. Sharkey TD, Seemann JR, Pearcy RW (1986) Contribution of metabolites of photosynthesis to postillumination CO_2 assimilation in response to lightflects. Plant Physiol 82: 1063–1068
7. Tcherkez G, Cornic G, Bligny R, Gout E, Ghashghaie J (2005) In vivo respiratory metabolism of illuminated leaves. Plant Physiol 138:1596–1606
8. Cournac L, Dimon B, Peltier G (1993) Evidence for ^{18}O labeling of photorespiratory CO_2 in photoautotrophic cell cultures of higher plants illuminated in the presence of $^{18}O_2$. Planta 190:407–414
9. Atkin OK, Evans JR, Siebke K (1998) Relationship between the inhibition of leaf respiration by light and enhancement of leaf dark respiration following light treatment. Aust J Plant Physiol 25:437–443
10. Valentini R, Epron D, de Angelis P, Matteucci G, Dreyer E (1995) In situ estimation of net CO_2 assimilation, photosynthetic electron flow and photorespiration in Turkey oak (Q. cerris L.) leaves: diurnal cycles under different levels of water supply. Plant Cell Environ 18:631–640
11. Abadie C, Boex-Fontvieille ERA, Carroll AJ, Tcherkez G (2016) In vivo stoichiometry of photorespiratory metabolism. Nat Plants 2: 15220
12. Gauthier PPG, Battle MO, Griffin KL, Bender ML (2018) Measurement of gross photosynthesis, respiration in the light, and mesophyll conductance using $H_2{}^{18}O$ labeling. Plant Physiol 177:62–74
13. Berry JA, Osmond CB, Lorimer GH (1978) Fixation of $^{18}O_2$ during photorespiration: kinetic and steady-state studies of the photorespiratory carbon oxidation cycle with intact leaves and isolated chloroplasts of C_3 plants. Plant Physiol 62:954–967
14. Abadie C, Tcherkez G (2019) Plant sulphur metabolism is stimulated by photorespiration. Commun Biol 2:379
15. Abadie C, Bathellier C, Tcherkez G (2018) Carbon allocation to major metabolites in illuminated leaves is not just proportional to photosynthesis when gaseous conditions (CO_2 and O_2) vary. New Phytol 218:94–106
16. Abadie C, Lalande J, Limami AM, Tcherkez G (2021) Non-targeted ^{13}C metabolite analysis demonstrates broad re-orchestration of leaf metabolism when gas exchange conditions vary. Plant Cell Environ 44:3171–3183
17. Abadie C, Lothier J, Boex-Fontvieille E, Carroll A, Tcherkez G (2017) Direct assessment of the metabolic origin of carbon atoms in glutamate from illuminated leaves using ^{13}C-NMR. New Phytol 216:1079–1089
18. Abadie C, Tcherkez G (2019) In vivo phosphoenolpyruvate carboxylase activity is controlled by CO_2 and O_2 mole fractions and represents a major flux at high photorespiration rates. New Phytol 221:1843–1852
19. Abadie C, Tcherkez G (2021) ^{13}C isotope labeling to follow the flux of photorespiratory intermediates. Plan Theory 10:427
20. Abadie C, Blanchet S, Carroll A, Tcherkez G (2017) Metabolomics analysis of postphotosynthetic effects of gaseous O_2 on primary metabolism in illuminated leaves. Funct Plant Biol 44:929–940
21. Tcherkez G, Mahé A, Guérard F, Boex-Fontvieille ERA, Gout E, Lamothe M, Barbour MM, Bligny R (2012) Short-term effects of CO_2 and O_2 on citrate metabolism in illuminated leaves. Plant Cell Environ 35:2208–2220
22. Aubert S, Gout E, Bligny R, Douce R (1994) Multiple effects of glycerol on plant cell metabolism. Phosphorus-31 nuclear magnetic resonance studies. J Biol Chem 269:21420–21427
23. Abadie C, Lalande J, Tcherkez G (2022) Exact mass GC-MS analysis: protocol, database, advantages and application to plant metabolic profiling. Plant Cell Environ 45:3171–3183

Chapter 17

Combining Gas Exchange and Rapid Quenching of Leaf Tissue for Mass Spectrometry Analysis Directly in Gas Exchange Cuvette

Yuan Xu, Xinyu Fu, Thomas D. Sharkey, and Berkley J. Walker

Abstract

Isotopically nonstationary metabolic flux analysis (INST-MFA) is a powerful technique for studying plant central metabolism, which involves introducing a $^{13}CO_2$ tracer to plant leaves and sampling the labeled metabolic intermediates during the transient period before reaching an isotopic steady state. The metabolic intermediates involved in the C_3 cycle have exceptionally fast turnover rates, with some intermediates turning over many times a second. As a result, it is necessary to rapidly introduce the label and then rapidly quench the plant tissue to determine concentrations in the light or capture the labeling kinetics of these intermediates at early labeling time points. Here, we describe a rapid quenching (0.1–0.5 s) system for $^{13}CO_2$ labeling experiments in plant leaves to minimize metabolic changes during labeling and quenching experiments. This system is integrated into a commercially available gas exchange analyzer to measure initial rates of gas exchange, precisely control ambient conditions, and monitor the conversion from $^{12}CO_2$ to $^{13}CO_2$.

Key words Rapid quenching, Gas exchange cuvette, Leaf tissue sampling, $^{13}CO_2$ labeling system, Rapid switching, Mass spectrometry.

1 Introduction

In autotrophic plant tissues such as leaves, $^{13}CO_2$ labeling is becoming more commonly used for introducing ^{13}C into central metabolism. During a $^{13}CO_2$ labeling experiment, the labeled carbon is initially incorporated into the C_3 cycle and photorespiration, and subsequently distributed to other metabolic pathways over time. As a result, intermediate metabolite pools in these pathways become labeled, reflecting the kinetics of carbon incorporation and flux through the metabolic network. Unlike other isotope tracing techniques, the steady-state $^{13}CO_2$ labeling in plants does not contain flux information, but the isotopically nonsteady state kinet ics do. This nonsteady-state condition is a critical factor for accurate

Berkley J. Walker (ed.), *Photorespiration: Methods and Protocols*, Methods in Molecular Biology, vol. 2792,
https://doi.org/10.1007/978-1-0716-3802-6_17,

data interpretation and flux modeling, necessitating the use of sophisticated mathematical models such as isotopically nonstationary metabolic flux analysis (INST-MFA) [1]. In recent years, the development of modeling techniques like INST-MFA has significantly expanded the utility and benefits of $^{13}CO_2$ labeling experiments in plant metabolism research [2–5].

INST-MFA provides a broad platform for quantifying metabolic fluxes through plant central metabolism using the transient labeling technique of ^{13}C isotope incorporation. By collecting samples at multiple time points during the labeling period, researchers can measure the labeling kinetics of different metabolic intermediates using mass spectrometry and use this information to estimate metabolic fluxes through the network. To ensure accurate measurement and interpretation of metabolic fluxes, several key experimental design considerations should be considered. These include (1) maintaining a pseudo-steady state of plant metabolism throughout the entire ^{13}C labeling period; (2) selecting appropriate time points to best capture the dynamic labeling process; (3) rapidly switching from $^{12}CO_2$ to $^{13}CO_2$ in the cuvette at the beginning of labeling process; and (4) rapidly quenching tissues to minimize metabolic artifacts from turnover of the labeling process. To address these experimental design considerations, it is necessary to have a system that can precisely control the environmental conditions, rapidly switch from $^{12}CO_2$ to $^{13}CO_2$, and facilitate rapid quenching of samples.

Rapid quenching is particularly important for analyzing C_3 cycle metabolic intermediates, which have exceptionally quick turnover rates between 0.08 and 3.23 s [6]. The rapid turnover of these intermediates makes rapid quenching crucial for accurate metabolic concentration and degree of ^{13}C labeling (mass isotopologue distribution) measurements, as even brief periods of darkness can significantly affect fluxes through these intermediates. In INST-MFA, it is crucial to sample C_3 cycle intermediates at early labeling time points (from a few seconds to a few minutes), when labeling patterns change rapidly, and to minimize the delay between the onset of labeling and quenching to avoid confounding effects due to metabolic changes during sampling. To guarantee that the labeling data are representative of in vivo conditions, both the switching time from $^{12}CO_2$ to $^{13}CO_2$ and the quenching time must be as fast as possible.

The quenching speed, or time it took to reach certain temperatures necessary to halt biochemical reactions, of our system was 0.1 s to reach −10 °C at the leaf center and 0.5 s to reach −10 °C at the leaf margin area enclosed at the edge of the gas exchange cuvette. This was measured with a datalogger and a butt-joint fine wire (80 μM) thermocouple threaded through a leaf [3]. To compare the effectiveness of our improved fast quenching system with the previous slow quenching method, which involves removing the

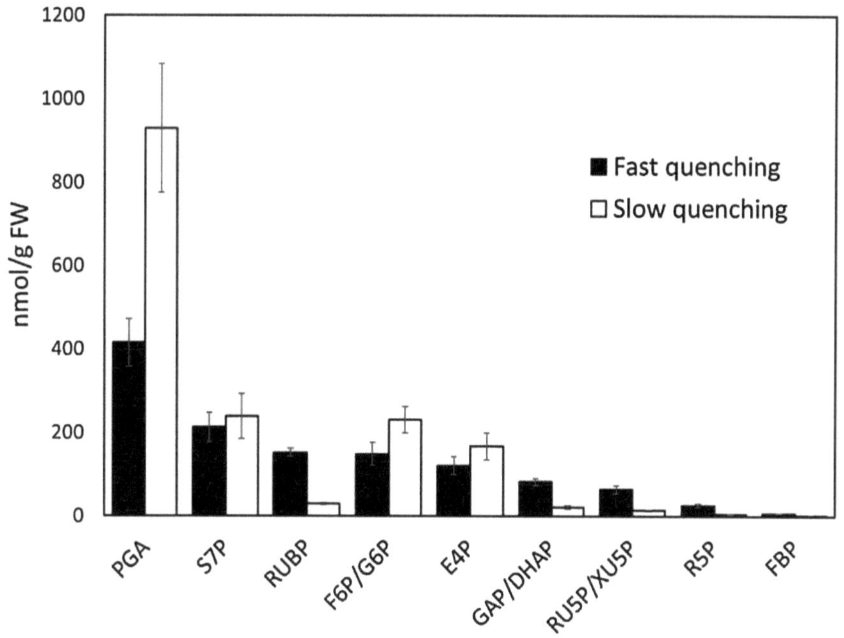

Fig. 1 The levels of C_3 cycle intermediates with slow and fast quenching sampling. PGA 3-Phosphoglyceric acid, S7P Sedoheptulose 7-phosphate, RUBP Ribulose 1,5-bisphosphate, F6P Fructose 6-phosphate, G6P Glucose 6-phosphate, E4P Erythrose 4-phosphate, GAP Glyceraldehyde 3-phosphate, DHAP Dihydroxyacetone phosphate, Ru5P ribulose-5-phosphate, X5P Xylulose 5-phosphate, R5P Ribose 5-phosphate, FBP Fructose 1,6-bisphosphatase

leaf from the cuvette and dipping it into liquid nitrogen for around 10 s, we measured the levels of C_3 cycle intermediates. The slow and fast quenching methods resulted in significantly different levels of metabolites measured (Fig. 1). The levels of ribulose 1,5-bisphosphate, ribose 5-phosphate, ribulose-5-phosphate/xylulose 5-phosphate, and glyceraldehyde 3-phosphate/dihydroxyacetone phosphate were found to be higher in the fast-quenching method. These results demonstrate that metabolic concentration is strongly influenced by the quenching speed. The fast turnover of the C_3 cycle intermediates can cause changes during slow quenching, and it is likely that the conversion from ribulose 1,5-bisphosphate to 3-phosphoglyceric acid occurred during the time the leaf was out of illumination before being dipped into the liquid nitrogen.

In this chapter, we describe the application of $^{13}CO_2$ feeding for labeling photosynthetic tissue using gas exchange and rapid quenching to perform INST-MFA in leaf metabolism. To rapidly quench metabolites in 0.5 s or less, we spray liquid nitrogen directly onto the leaf in a gas exchange cuvette. Combining gas exchange and rapid quenching directly in a gas exchange cuvette provides more precise control over the environmental conditions during the labeling and quenching process. Additionally, it enables the monitoring of the leaf's net CO_2 assimilation rate before labeling, ensuring that the metabolic state of the leaves is comparable and providing an additional flux constraint for INST-MFA analysis.

2 Materials

1. Infra-red gas analyzer (IRGA) (*see* **Note 1**).
2. $^{12}CO_2$ cartridge, soda lime, and desiccant for the IRGA.
3. Custom-modified 9 cm^2 leaf cuvette for the IRGA with an injection borehole into the lower half of the cuvette.
4. Mount for the IRGA measurement head that can hold it perpendicularly to minimize contact between liquid nitrogen and internal components of the IRGA.
5. Four-way valve to toggle the CO_2 supply from the IRGA console and the mixed air containing $^{13}CO_2$.
6. Mass flow controllers (for O_2 and N_2 a range of 1–5 standard Liters per minute and for the $^{13}CO_2$ supply a range of 30 standard cubic centimeters per minute are good ranges).
7. Tubing and compression fittings.
8. Nitrogen gas tank with two-stage regulator.
9. Oxygen gas tank with two-stage regulator.
10. $^{13}CO_2$ (isotopic purity 99 atom percent) gas tank with two-stage regulator diluted to ~13% of the total pressure within the lecture bottle with N_2 or CO_2-free gas.
11. Water bath.
12. Impinger for humidifying incoming tank air with moistened Stuttgarter Masse (this reduces the time it takes for the $13CO_2$ to come into equilibrium with the humidifying substrate).
13. Liquid nitrogen sprayer (*see* **Note 2**).
14. Liquid nitrogen.
15. 1.5 mL microcentrifuge tubes.
16. Controller for the mass flow controllers (*see* **Note 3**).
17. Putty for sealing hole used for spraying liquid nitrogen.

3 Methods

3.1 Instrument Setup

1. Use tubing and fittings to connect mass flow controllers with external O_2, N_2, and $^{13}CO_2$ tanks (Fig. 2).
2. Connect mass flow controllers to control unit.

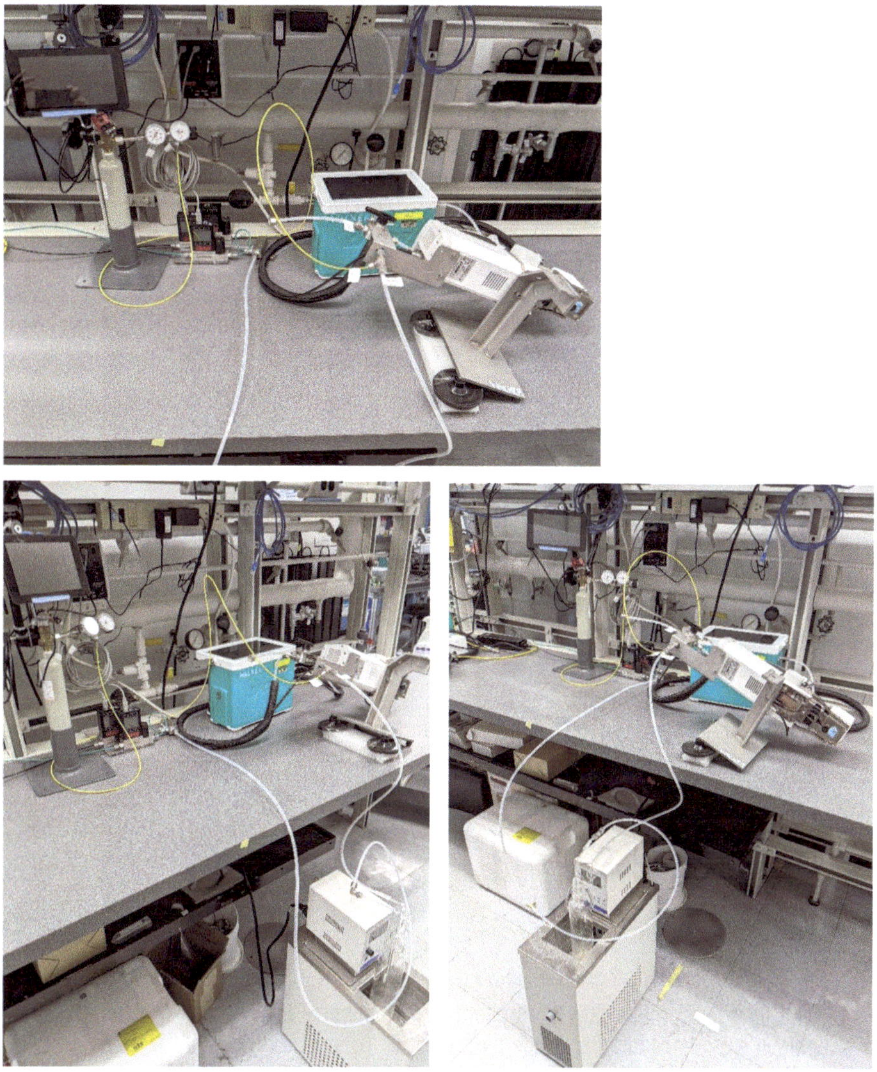

Fig. 2 $^{13}CO_2$ labeling system combining gas exchange and rapid quenching. The system is comprised a portable IRGA photosynthesis system, a custom leaf cuvette with liquid nitrogen injection port, mass flow controllers, a Raspberry Pi computer with a monitor, and gas tanks. The tubing colors reflect the gas compositions used with green representing oxygen, black nitrogen, yellow carbon dioxide, and clear for a mixed gas stream

3. Pass the airflow from the mass flow controllers to a water bath with the impinger to humidify the O_2 and N_2 mixture before mixing with the $^{13}CO_2$-labeled gas (Fig. 3) (*see* **Note 4**).
4. Set the water bath temperature so it provides a similar amount of humidity as the incoming air to the leaf in the IRGA. This is important to minimize the difference in vapor pressure difference (VPD) between the leaf and the incoming gas.

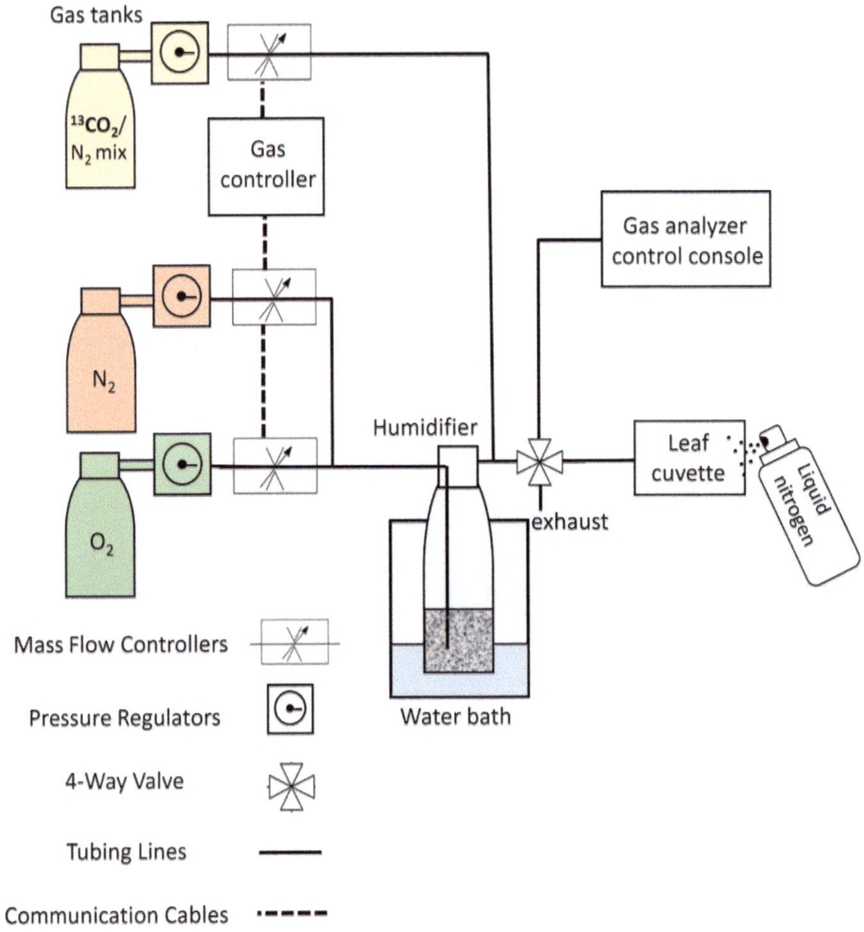

Fig. 3 Mixed gas flow path in the $^{13}CO_2$ labeling systems. A gas mixture comprising 78% N_2, 21% O_2, and 420 ppm of $^{13}CO_2$ is set by mass flow controllers and Raspberry Pi. The gas mixture passes through a humidifier to achieve a similar humidity in the leaf cuvette. To start the transient labeling process, turn the 4-ways valve to introduce the mixed gas into the IRGA and leaf cuvette. At the designated time intervals, remove the adhesive putty and spray liquid nitrogen into the leaf cuvette to quench metabolism

5. To minimize the time between switching to the $^{13}CO_2$ background and the label reaching the leaf, connect the four-way valve as close to the measurement cuvette as possible. The initial state of this valve should pass the air pumped from the IRGA console into the measurement cuvette, while the incoming labeled air is vented. When the valve is toggled, it will vent the console air and pass the labeled air into the cuvette (*see* **Note 5**).
6. Modify a 9 cm^2 cuvette for an IRGA by adding a hole on the lower edge of the cuvette to accommodate a liquid nitrogen cryospray nozzle.

7. Use a stopper of adhesive poster putty to seal the injection port of the leaf cuvette during the initial gas exchange and labeling period.
8. Check that the pressure of the cuvette is positive to ensure that any leaks during this time are from inside to outside the cuvette and will not affect the internal gas concentration (*see* **Note 6**).

3.2 Preparing $^{13}CO_2$ Gas for Labeling Experiments

1. Dilute the $^{13}CO_2$ tank with N_2 to ~13% according to pressure (*see* **Note 7**).
2. Turn on mass flow controllers and gas tanks.
3. Test the control of the mass flow controllers for N_2, O_2, and $^{13}CO_2$ using the Raspberry Pi monitor (or other device to control the mass flow controllers).
4. Set N_2 to 78% and O_2 to 21% and pass them through the impinger humidifier placed in a water bath (Fig. 3).
5. Calculate the flow rate needed to set the concentration of $^{13}CO_2$ at 40 μL L^{-1} or the initial concentration dictated by the research aims of the labeling time course.

3.3 Gas Exchange Measurements

1. Turn on IRGA and run system warmup tests to ensure the functioning of the environment control and sensors.
2. Set the desired cuvette condition (e.g., CO_2 concentration, light intensity, humidity, and temperature) that is close to the growth environment of the plants or is in line with the research aims of the labeling time course. Set the chamber airflow to 420 μmol s^{-1} and the chamber mixing fan speed to 10,000 rpm.
3. Place a fully developed leaf in the measurement cuvette for 10–30 min to acclimate (*see* **Note 8**).
4. Record the net CO_2 assimilation rate after the leaf reaches a photosynthetic steady state.

3.4 $^{13}CO_2$ Labeling Experiments

1. Conduct the labeling experiment between the morning and early afternoon to align with the steady state for the metabolism of sucrose and starch [2].
2. Set the flow rate for the humidified mixture of 420 μL L^{-1} $^{13}CO_2$, 78% N_2, and 21% O_2 using three gas mass flow controllers to 5 L min^{-1} and hold for about 5 s to flush the tubing (*see* **Note 9**).
3. Change the IRGA inlet pump from "auto" to "manual."
4. Use the four-way valve on the leaf cuvette to switch the gas supply from the IRGA console to the humidified $^{13}CO_2$-containing gas mixture.
5. Start a timer to indicate the labeling duration.

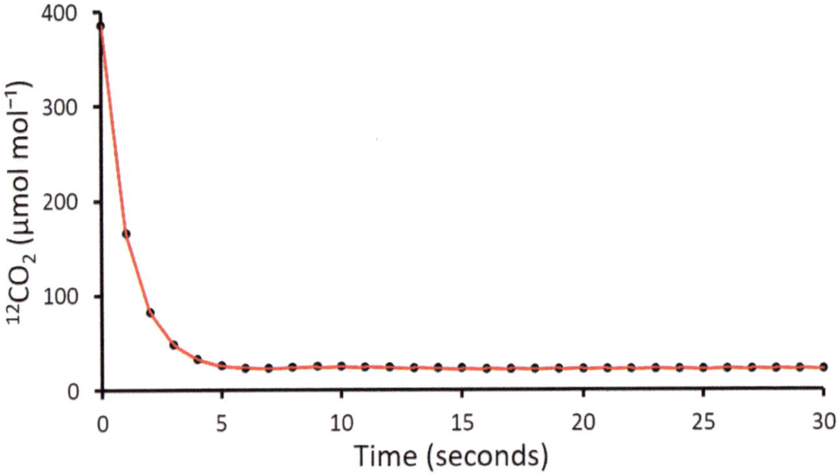

Fig. 4 Use the IRGA to monitor the $^{12}CO_2$ concentration in the cuvette and estimate the time required for the complete transition from $^{12}CO_2$ to $^{13}CO_2$. It is important to switch from $^{12}CO_2$ to $^{13}CO_2$ atmospheres quickly. From the start of the $^{13}CO_2$ labeling experiment, the IRGA should track the decline of $^{12}CO_2$ due to the insensitivity of the IRGA to $^{13}CO_2$. The $^{12}CO_2$ should be flushed out of the cuvette by 5 s when the flow rate of $^{13}CO_2$ reaches 5 L min^{-1}

6. Monitor the $^{12}CO_2$ concentration in the cuvette using the IRGA to estimate the gas transition time. The complete change from $^{12}CO_2$ to $^{13}CO_2$ will take approximately 5 s (Fig. 4) (*see* **Note 10**).
7. Once the atmosphere is switched to $^{13}CO_2$, switch the total flow rate from the mass flow controllers to 1 L min^{-1} (or whatever flow gives a similar sample flow as what was used by the IRGA) for the rest of the labeling period.
8. After the appropriate labeling period, the leaves are now ready for rapid quenching (*see* **Notes 11–14**).

3.5 Rapid Quenching

1. A few seconds before the desired labeling time, remove the adhesive putty from the leaf surface (*see* **Notes 6**).
2. At the exact labeling time, spray liquid nitrogen directly onto the leaf using the injection port on the cuvette to stop metabolism at a set time point (*see* **Notes 15** and **16**).
3. Quickly open the leaf cuvette and use scissors to cut off the frozen areas (indicated by white tissue) of the leaf as quickly as possible (*see* **Note** 17).
4. Quickly put the labeled sample in liquid nitrogen and transfer it to a 1.5 mL microcentrifuge tube (*see* **Note 18**).
5. Keep the sample at −80 °C for subsequent extraction and mass spectrometry analysis. We keep samples for up to 6 months, but longer or shorter may be appropriate depending on the metabolites to be analyzed and possibly the species of the plant being measured.

4 Notes

1. We have used this setup with a LI-6800 portable photosynthesis system (LI-COR Biosciences, Lincoln, NE, USA), but the principles may be applicable to other IRGA systems. The LI-6800 is internally fairly well-protected against over-spray of liquid nitrogen so any other IRGA should be trialed first with great caution. Within later notes, we outline some tips for minimizing the amount of liquid nitrogen that enters the instrument. In our experience, the only issues have been that the flow sensors need to be zeroed more frequently and that we will occasionally freeze and crack the rubber tubing/gasket that runs between the cuvette and the IRGA head. These can be replaced fairly easily and it helps to go ahead and buy a large amount for quick replacements.
2. These sprayers are commonly used in dermatology and there are a number of suppliers.
3. We utilize a custom-programmed touch screen-equipped Raspberry Pi to operate the mass flow controllers, enabling us to adjust both gas composition concentration and flow rate quickly during the experiment.
4. Set the water bath temperature so it provides a similar amount of humidity as the incoming air to the leaf in the IRGA. This is important to minimize the difference in water vapor difference between the leaf and the incoming gas.
5. Venting this air is important to prevent over-pressurizing the system.
6. It is possible to set and monitor the cuvette overpressure in the LI-6800. This overpressure is important for two reasons: (1) it is important to negate the consequences of leaks during the measurement and (2) to prevent unlabeled CO_2 from entering the cuvette during the spraying procedure once the putty is removed for spraying during the rapid quench.
7. Dilution of the $^{13}CO_2$ allows for the gas to be delivered at a higher flow rate into the N_2/O_2 mix, which speeds the time it takes to reach a set CO_2 concentration and also allows most of the $^{13}CO_2$ to be removed from the bottle once ambient pressure is reached. This dilution can be done using various gas mixing systems based on pressure. Since high pressures are involved in this mixing and care is needed not to exceed the limits of the lecture bottle, this should be done by those who are familiar with this approach and have the proper setup to do so. If this equipment and expertise are not available, the N_2, O_2, and $^{13}CO_2$ mixing systems can be replaced by a custom tank of $^{13}CO_2$ gas.

8. During transient labeling, it is important to maintain a steady state of plant metabolism. A metabolic steady state will take longer to reach than a photosynthetic steady state. One way to establish if a system is in a metabolic steady state during the labeling is to perform a preliminary time course with no label and then to check that all the metabolites have a similar abundance at all time points. Once a labeling experiment is done, the relative abundance of each metabolite can also be compared to ensure that there are no trends of increasing or decreasing pool sizes. In our experience with the species, we have labeled (Arabidopsis and Tobacco) this takes about 10–30 min.
9. An initial high flow rate is important to quickly introduce $^{13}CO_2$ and displace any remaining $^{12}CO_2$.
10. Since many IRGA CO_2 analyzers are insensitive to $^{13}CO_2$, the measured CO_2 concentration in the IRGA will drop when transitioning to $^{13}CO_2$. This is a good indication of how fast the cuvette has transitioned to the labeled gas. This also means that any assimilation values measured under the labeled CO_2 are not correct.
11. In previous work, time points of 0, 0.5, 1, 2, 2.5, 3, 5, 7, 10, 15, 30, and 60 min were used, with the sequence of each time point assessed being randomized [3–5].
12. It is recommended to collect samples more frequently near the early time points to capture the kinetics of processes that interact with the CBC cycle.
13. It can be difficult to collect time points shorter than 5 s due to the 5 s washout time for $^{12}CO_2$ in the leaf cuvette (Fig. 4).
14. For some pools, additional lengthy time points (such as 1 h) are useful to obtain the active versus inactive pool sizes.
15. To avoid liquid nitrogen from damaging the rubber gasket on the IRGA during the spray, position the cuvette at a 30-degree angle so that the excess liquid nitrogen will not pass into the instrument. Similar considerations may be needed for other IRGA manufacturers.
16. The duration of the liquid nitrogen spray on the leaf needs to be tested to ensure it is long enough to quench the whole leaf but not too much to break the leaf.
17. We use the white area of the leaf to indicate that the leaf is still frozen. Importantly, the leaves are not allowed to thaw at any time prior to metabolite extraction.
18. If the leaf sample is too large to be directly placed into a 1.5 mL microcentrifuge tube, it can be crushed using liquid nitrogen in a Falcon tube and then carefully transferred to the microcentrifuge tube.

Acknowledgments

This work was supported by Office of Basic Energy Sciences of the United States Department of Energy (Grant DE-FG02-91ER20021 and Grant DE-FOA-0001650).

References

1. Young JD (2014) INCA: a computational platform for isotopically non-stationary metabolic flux analysis. Bioinformatics 30:1333–1335
2. Ma F, Jazmin LJ, Young JD, Allen DK (2014) Isotopically nonstationary 13C flux analysis of changes in Arabidopsis thaliana leaf metabolism due to high light acclimation. Proc Natl Acad Sci 111:16967–16972
3. Xu Y, Fu X, Sharkey TD, Shachar-Hill Y, Walker BJ (2021) The metabolic origins of non-photorespiratory CO2 release during photosynthesis: a metabolic flux analysis. Plant Physiol 186:297–314. https://doi.org/10.1093/plphys/kiab076
4. Xu Y, Wieloch T, Kaste JAM, Shachar-Hill Y, Sharkey TD (2022) Reimport of carbon from cytosolic and vacuolar sugar pools into the Calvin–Benson cycle explains photosynthesis labeling anomalies. Proc Natl Acad Sci 119: e2121531119
5. Fu X, Gregory LM, Weise SE, Walker BJ (2022) Integrated flux and pool size analysis in plant central metabolism reveals unique roles of glycine and serine during photorespiration. Nat Plants 9:1–10
6. Arrivault S, Guenther M, Ivakov A, Feil R, Vosloh D, Van Dongen JT, Sulpice R, Stitt M (2009) Use of reverse-phase liquid chromatography, linked to tandem mass spectrometry, to profile the Calvin cycle and other metabolic intermediates in Arabidopsis rosettes at different carbon dioxide concentrations. Plant J 59:826–839. https://doi.org/10.1111/j.1365-313X.2009.03902.x

Part IV

Computational Tools

Chapter 18

Reaction-Kinetic Modeling of Photorespiration Using Modelbase

Tim Nies and Marvin van Aalst

Abstract

Plant science has become more and more complex. With the introduction of new experimental techniques and technologies, it is now possible to explore the fine details of plant metabolism. Besides steady-state measurements often applied in gas-exchange or metabolomic analyses, new approaches, e.g., based on ^{13}C labeling, are now available to understand the changes in metabolic concentrations under fluctuating environmental conditions in the field or laboratory. To explore those transient phenomena of metabolite concentrations, kinetic models are a valuable tool. In this chapter, we describe ways to implement and build kinetic models of plant metabolism with the Python software package `modelbase`. As an example, we use a part of the photorespiratory pathway. Moreover, we show additional functionalities of `modelbase` that help to explore kinetic models and thus can reveal information about a biological system that is not easily accessible to experiments. In addition, we will point to extra information on the mathematical background of kinetic models to give an impetus for further self-study.

Key words Kinetic modeling, Modelbase, Model building, Simulations, Isotope labeling, Metabolic control analysis

1 Introduction

The combination of experimental and theoretical approaches for illuminating biological phenomena leads to a deeper understanding of complex systems. This approach has been successfully applied in fields such as gene regulatory networks, metabolic pathways, and specifically in the plant physiological context to explore intricate connections between the electron transport chain, Calvin Benson Bassham cycle, and photorespiration [1–4]. The description of biological phenomena with mathematical models has become a necessary and valuable tool in the repertoire of biological sciences leading to a wide variety of techniques to model a biological system

Tim Nies and Marvin van Aalst authors contributed equally to this work.

Berkley J. Walker (ed.), *Photorespiration: Methods and Protocols*, Methods in Molecular Biology, vol. 2792, https://doi.org/10.1007/978-1-0716-3802-6_18,

in silico, ranging from stochastic and kinetic to genome-scale metabolic models. This chapter will overview how kinetic models can be implemented in two ways using the programming language Python (**Note 1**). We will primarily focus on using the software package `modelbase` as an easily accessible computational tool for kinetic modeling with broad functionalities [5, 6].

Many other introductory texts about building mathematical models are not very specific and often address a broad audience with a deeper mathematical background, making the first steps for experimental plant biologists difficult. In contrast to this, we focus in this chapter on photorespiration-related examples that are known to all scientists working in plant biology. Moreover, the content aims to enable the reader to implement simple models in Python quickly, thus keeping optional mathematical details short. The chapter starts with setting up the computer to be ready for modeling and follows in its structure an introductory course in mathematical modeling of biological systems. Especially in a book based on protocols and methods, this strategy is advantageous because it guides the reader through the whole process, from setting up the computer to plotting the first simulations.

Kinetic models use differential equations (**Note 2**) as the mathematical backbone, a basic introduction to which can be found in the textbook by Hiroki Sayama (https://bingweb.binghamton.edu/~sayama/textbook/, last accessed 05 December 2022).

A model focused on metabolic processes consists of three parts, *see* Fig. 1

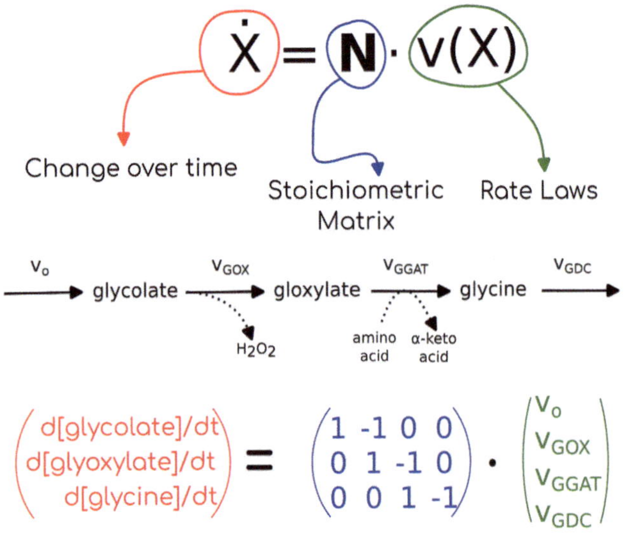

Fig. 1 The general concept of kinetic models and scheme of the extract of the photorespiratory pathway that is used as a toy model. *GOX* glycolate oxidase, *GGAT* glutamate:glyoxylate aminotransferase, *GDC* glycine decarboxylation complex

1. The derivatives, describing the change of a system variable, such as metabolite concentrations, over time.
2. The stoichiometric matrix **N**, encoding the underlying structure of the metabolic network.
3. Rate equations, which can be found in many forms, like mass-action kinetics, Michaelis–Menten kinetics, random or ordered Uni-Uni, Uni-Bi, Bi-Uni, Bi-Bi, or Bi-Bi Ping Pong mechanisms. An excellent introduction to this topic is the articles by Cleland or this book by Cornish-Bowden [7, 8]. Rate equations are often derived using the King-Altman algorithm. This paper [9] explains the algorithm in more depth than it is typically presented.

The rest of the chapter is structured in the following way. First, we describe one of many possible computer setups to start modeling. Next, we will show two ways to implement a reaction sequence (GOX, GGAT, and GDC) extracted from a published kinetic model of photorespiration [4]. This will be the toy model with which most examples guide the reader through model building. Lastly, additional functionalities of the software package `modelbase` will be explored.

2 Computer Setup

A computational tool is necessary to start modeling that translates a mathematical description of a kinetic model (system of differential equations) into a computational executable format. For this, scientists use computer languages like C, C++, Julia, R, Rust, Matlab, or Python. Often one language is preferable to another in a different scientific context like the scientific field and current standards. Here we use Python to implement and build a kinetic model. We will install Python using the mamba package manager (https://github.com/conda-forge/miniforge#mambaforge), (**Note 3**).

1. Download mamba from https://github.com/conda-forge/miniforge#mambaforge (last accessed 05 December 2022) and follow the installation instructions.
2. Open up a shell (e.g., powershell for windows and bash for unix) and create a virtual environment to isolate the packages installed for the examples in this chapter

```
$ mamba create -n tutorial python=3.11 modelbase jupyter
```

3. Activate the virtual environment

```
$ mamba activate tutorial
```

4. Open up a jupyter notebook server

```
(tutorial) $ jupyter notebook
```

You are ready! The model-building process can begin.

3 Building and Simulating Models Using SciPy

One of the most basic ways to implement a kinetic model described with ordinary differential equations is by handwriting it, which is illustrative for small models, but becomes cumbersome and hard to follow for more complex models. For completeness, we will show you how to implement the toy model (the GOX, GGAT, and GDC catalyzed reaction sequence) depicted in Fig. 1, by hand and contrast it with the `modelbase` implementation in the subsequent sections. Note this toy model is an extraction of the photorespiratory pathway and thus is limited in its prediction capacity. According to the model in Fig. 1, the differential equations (**Note 4**) describing the change of glycolate, glyoxylate, and glycine over time are

$$
\begin{aligned}
\frac{d[\text{glycolate}]}{dt} &= v_o - v_{\text{GOX}} \\
\frac{d[\text{glyoxylate}]}{dt} &= v_{\text{GOX}} - v_{\text{GGAT}} \\
\frac{d[\text{glycine}]}{dt} &= v_{\text{GGAT}} - v_{GDC}
\end{aligned}
$$

As done in [4], let us assume that the rate equations v_{GOX}, v_{GGAT}, and v_{GDC} are of Michaelis–Menten form (glutamate for the GGAT-reaction is assumed to be in excess). The influx v_o is treated as a constant. If we want to implement and solve these equations, we need

1. Parameters for the rate equations
2. A differential equation solver, **Note 5**
3. Initial conditions for the system variables, e.g., metabolite concentrations, **Note 6**

Let us begin by encoding our equation system in Python. First we import the required packages and define the parameters. Note how the parameters are specified as a dictionary to easily identify them later in the code.

```python
import matplotlib.pyplot as plt
import numpy as np
from scipy.integrate import solve_ivp
from typing import Iterable

pars = {"Vm1": 100 * 4, "Km1": 0.38, "Vm2": 95 * 4, "Km2": 1.2,
        "Vm3": 100 * 4, "Km3": 1.0,  "v0": 60 * 4,}
```

Next we define a Michaelis–Menten type rate law

```
def mm(S: float, Vm: float, Km: float) -> float:
    return S * Vm / (S + Km)
```

and set up our system of ordinary differential equations. Note that the function takes both the time t as well as a vector y as inputs, even though the system does not explicitly use the time t. This is because the `solve_ivp` function requires our initial value problem to be of the shape `dy / dt = f(t, y)`.

```
def system(
    t: float, y: Iterable[float], p: dict[str, float]
) -> Iterable[float]:
    glycolate, glyoxylate, glycine = y
    # v(X)
    v0 = p["v0"]
    vgox = mm(glycolate, p["Vm1"], p["Km1"])
    vggat = mm(glyoxylate, p["Vm2"], p["Km2"])
    vgdc = mm(glycine, p["Vm3"], p["Km3"])
    # N * v(X)
    dglycolate_dt = v0 - vgox
    dglyoxylate_dt = vgox - vggat
    dglycine_dt = vggat - vgdc
    return dglycolate_dt, dglyoxylate_dt, dglycine_dt
```

Lastly we integrate the system over time using the `solve_ivp` function from the scipy package (https://scipy.org/, last accessed 05 December 2022), which takes the system, initial conditions, and time points as inputs and returns the solution between the specified time points. Because we assume a constant influx into the system, we can set all initial values to zero. The time points of interest we define as ranging from 0 to 0.5 hours (30 minutes). The choice of the time in which we observe the output of the toy model is arbitrary and was selected as shown because the model approaches quickly a steady state. However, when modeling more realistic biological processes, the time of the simulation should be in a range where the whole model behavior can be observed. Note that `solve_ivp` requires any additional args to x and t to be passed as a tuple.

```
sol = solve_ivp(
    system,             # function
    t_span=(0, 0.5),    # time points
    y0=[0, 0, 0],       # initial conditions
    args=(pars,),       # parameters
)
```

We can take the solution of `solve_ivp` and plot via the Python package matplotlib (https://matplotlib.org/, last accessed 05 December 2022).

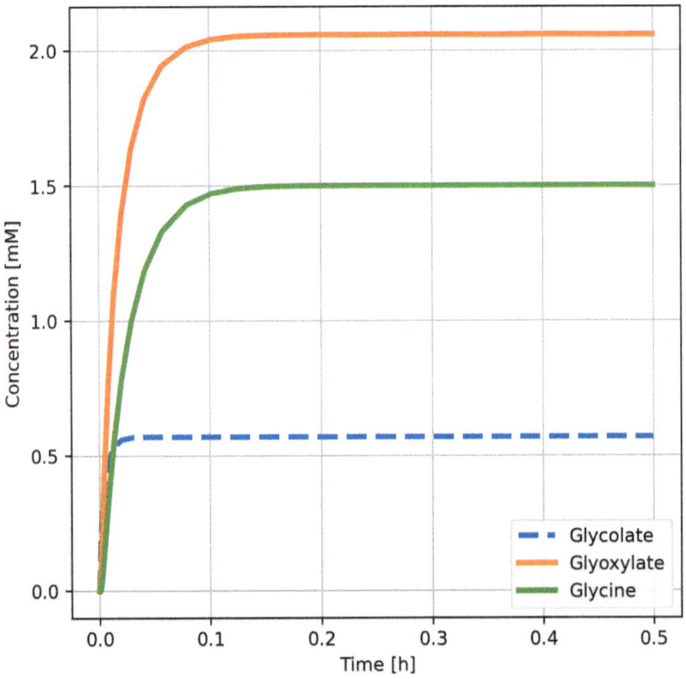

Fig. 2 Simulation output of the photorespiratory toy model consisting of the steps catalyzed by GOX, GGAT, and GDC. Shown is a time course of the metabolite concentrations over a time range of 30 minutes

```
fig, ax = plt.subplots(figsize=(6, 6))
ax.plot(sol.t, sol.y[0], lw=3, ls="--", label="Glycolate")
ax.plot(sol.t, sol.y[1], lw=3, label="Glyoxylate")
ax.plot(sol.t, sol.y[2], lw=3, label="Glycine")

# add labels, a legend and a grid
ax.set_xlabel("Time [h]")
ax.set_ylabel("Concentration [mM]")
ax.legend(loc="lower right")
ax.grid()
```

The resulting plot is shown in Fig. 2.

4 Building Models with `modelbase`

Especially in the case of bigger models, the procedure, as mentioned above, becomes cumbersome. Difficulties that can be encountered when building big models without extra software are: problems to track changes and thus debugging possible errors, multiple variables with the same name, overwriting parts of the model without being aware of it, and much effort to extract information from the simulation output as well as to adapt the existing code. Thus multiple computer programs were developed to facilitate modeling, one of them being `modelbase`, which was designed to be easy to learn. An ordinary differential equation solver is already included in the software package. The first step in model building with `modelbase` is to define the model using parameters (constants) and compounds (variables), in this case glycolate, glyoxylate, and glycine (note that we are re-using the definitions from the scipy example).

```
from modelbase.ode import Model, Simulator


def constant(x: float) -> float:
    return x


m = Model(
    parameters=pars,
    compounds=["Glycolate", "Glyoxylate", "Glycine"],
)
```

Now we can add reactions to the model, for which we need to specify the following:

1. A memorable name with which we can identify the reaction
2. The rate function
3. The stoichiometry of the reaction
4. Arguments for the rate function

For our toy model with the four reactions and three metabolic compounds, we have to add the following lines to our model implementation:

```
m.add_reaction_from_args(
    rate_name="vo",
    function=constant,
    stoichiometry={"Glycolate": 1},
    args=["vO"],
)
m.add_reaction_from_args(
    rate_name="vGOX",
    function=mm,
    stoichiometry={"Glycolate": -1, "Glyoxylate": 1},
    args=["Glycolate", "Vm1", "Km1"],
)
m.add_reaction_from_args(
    rate_name="vGGAT",
    function=mm,
    stoichiometry={"Glyoxylate": -1, "Glycine": 1},
    args=["Glyoxylate", "Vm2", "Km2"],
)
m.add_reaction_from_args(
    rate_name="vGDC",
    function=mm,
    stoichiometry={"Glycine": -1},
    args=["Glycine", "Vm3", "Km3"],
)
```

Compared to building the model by hand using modelbase might seem more complicated. However, modelbase forces you to structure your model in an easily readable way and helps to keep track of all involved variables. Moreover, the stoichiometric matrix is automatically built by specifying the stoichiometry for each reaction. This functionality is helpful if you want to examine your kinetic model structurally.

We are ready with our model. Now we will give the Model object to the Simulator object. This object handles all simulations. We only need to specify the endpoint of our simulations.

```
s = Simulator(m)
s.initialise({"Glycolate": 0, "Glyoxylate": 0, "Glycine": 0})
_ = s.simulate(0.5)
fig, ax = s.plot(xlabel="Time [h]", ylabel="Concentration [mM]")
```

modelbase has many functionalities; please look at the official documentation (https://qtb-hhu.gitlab.io/modelbase-software/, last accessed 05 December 2022) for extensive information. As an example, we can change the influx function in the above model to a time-varying sine function (**Note 7**)

```
def sin(t: float) -> float:
    return 60 * 4 + 20 * np.sin(2 * np.pi * t * 0.1)


m.update_reaction_from_args(
    rate_name="vo",
    function=sin,
    args=["time"],
)
```

and again use the same pattern to integrate the model

```
s = Simulator(m)
s.initialise({"Glycolate": 0, "Glyoxylate": 0, "Glycine": 0})
s.simulate(24)
fig, ax = s.plot(xlabel="Time [h]", ylabel="Concentration [mM]")
```

The resulting plot of this non-autonomous ordinary differential equation system is shown in Fig. 3. Another useful functionality that finds application in everyday modeling is adding derived parameters to a model. A typical example is the recurring multiplication of the ideal gas constant and temperature in thermodynamic calculations. For the implementation of a derived parameter, we use the add_derived_parameter method.

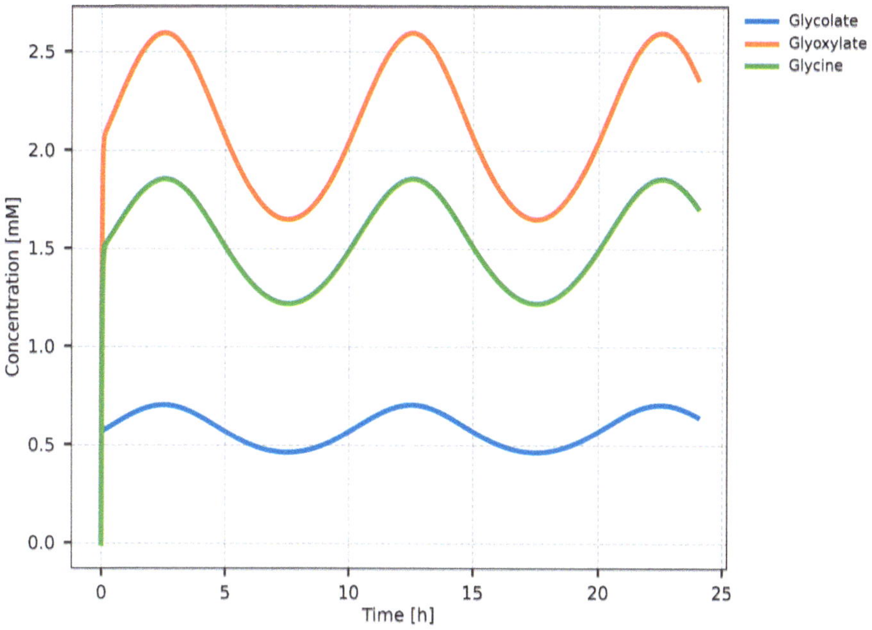

Fig. 3 Simulation output of the nonautonomous version of the photorespiratory toy model consisting of the steps catalyzed by GOX, GGAT, and GDC with a sine wave as influx. Shown is a time course of the metabolite concentrations over a time range of 24 hours

```
def proportional(x: float, y: float) -> float:
    return x * y


m = Model()
m.add_parameters(
    {
        "R": 8.314,  # Gas constant / (J / (K * mol))
        "T": 298.3,  # Temperature / K
    }
)
m.add_derived_parameter(
    parameter_name="RT",
    function=proportional,
    parameters=["R", "T"],
)
print(m.parameters["RT"])  # prints 2480.0662
```

5 Algebraic Modules

In many dynamic biological systems, we find constant quantities not influenced by the change of the system variables / metabolic concentrations. Examples are the total amount of adenosine-phosphates in glycolysis or the number of phosphate-containing compounds in the Calvin Benson Bassham cycle. These quantities are encoded in so-called conservation relations. It is possible to extract information about conservation relations and the steady-state fluxes by calculating the left and right nullspace of the stoichiometric matrix. For this, it is recommended to use computer

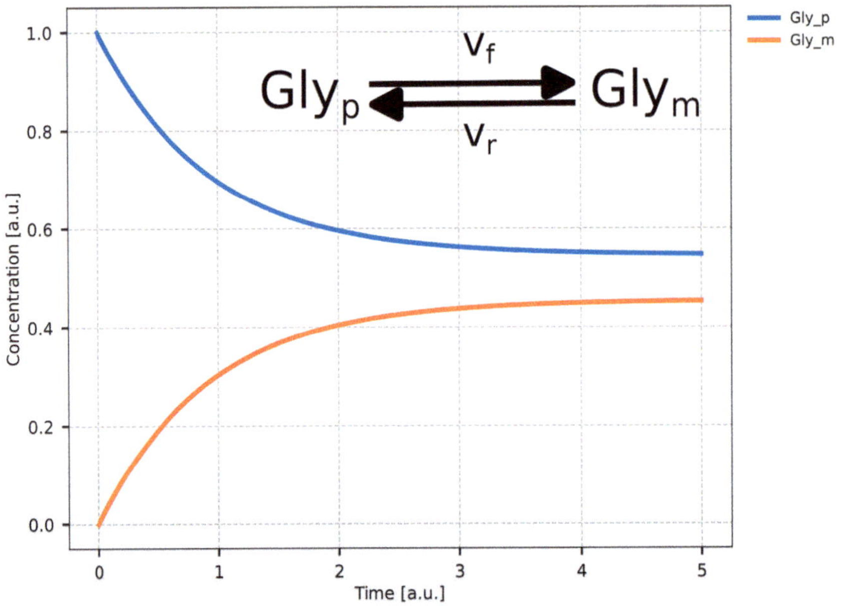

Fig. 4 Toy model (inlet) and simulation output over a time range of 5 time units demonstrating the usage of the algebraic module in `modelbase`. Gly_p—glycine in peroxisome and Gly_m—glycine in mitochondrion

algebra systems (CAS). In Python, the package sympy aims to become a completely functional CAS and can be used for small stoichiometric matrices [10]. Assuming we have identified such a conserved quantity, we can simplify the underlying differential equation system using the algebraic module in `modelbase`. The algebraic module can implement any algebraic equation, not necessarily a conservation relation, e.g., an equation of fluorescence yield in a description of the chloroplast-located electron transport chain. To demonstrate the algebraic module, we will use a different toy example. Assume we measure that the total pool size of glycine in a photosynthetic cell is constant over a large period. The total pool size is normalized to one. We observe that the distribution of glycine between the peroxisome (Gly_p) and mitochondrion (Gly_m) changes considerably due to photorespiratory activity and subsequent transport. Moreover, we assume that the space between the peroxisome and mitochondrion is small. This assumption allows us to ignore more complex spatial information. A mass-action kinetic describes the change in the distribution of glycine between the two cellular compartments (*see* Fig. 4 for the toy model and simulation output). The differential equation system for this example is

$$\frac{d[Gly_p]}{dt} = k_r[Gly_m] - k_f[Gly_p]$$
$$\frac{d[Gly_m]}{dt} = k_f[Gly_p] - k_r[Gly_m]$$

Because the sum of the system variables $[Gly_p]$ and $[Gly_m]$ is constant, $[Gly_p]+[Gly_m]=[Gly_{tot}]$, it is possible to simplify the system by using the fact that,

$$[Gly_m] = [Gly_{tot}] - [Gly_p]$$

In particular, this means we do not need to simulate $[Gly_m]$ because it can be derived at each time point by using the above relation. The algebraic module allows us to rewrite the equation system into differential-algebraic form. For this we use the `add_algebraic_module_from_args` method that requires to specify

1. The name of the algebraic module
2. The compound that is derived by the algebraic relation
3. The algebraic relation in form of a Python function
4. The arguments of the Python function

Thus the model implementation looks like

```
def moiety(amount: float, total: float) -> float:
    return total - amount


m = Model({"Gly_tot": 1, "kf": 0.5, "kr": 0.6})
m.add_compounds(["Gly_p"])

m.add_algebraic_module_from_args(
    module_name="Gly_m_module",
    derived_compounds=["Gly_m"],
    function=moiety,
    args=["Gly_p", "Gly_tot"],
)

m.add_reaction_from_args(
    rate_name="vf",
    function=lambda X1, kf: kf * X1,
    stoichiometry={"Gly_p": -1},
    args=["Gly_p", "kf"],
)

m.add_reaction_from_args(
    rate_name="vr",
    function=lambda X2, kr: kr * X2,
    stoichiometry={"Gly_p": 1},
    args=["Gly_m", "kr"],
)

s = Simulator(m)
s.initialise({"Gly_p": 1.0})
s.simulate(5)
fig, ax = s.plot_all(
    xlabel="Time [a.u.]", ylabel="Concentration [a.u.]"
)
```

6 Metabolic Control Analysis

Sensitivity analysis provides a theoretical foundation to systematically quantify the effects of small parameter perturbations on global system behavior [11, 12]. In particular, Metabolic Control Analysis (MCA), initially developed to study metabolic systems, is an important and widely used framework providing quantitative information about the response of the system to perturbations [13, 14], *see* **Note 8**. `modelbase` has a full suite of methods to calculate elasticities and response coefficients, which can be plotted as a heatmap, giving a clear and intuitive color-coded visualization of the results. For example, to calculate the response coefficients of multiple parameters, we can use the `get_response_coefficients_df` function in the `mca` module, which conveniently returns both the concentration and flux response coefficients as separate pandas data frames. This function requires us to specify which parameters shall be perturbed and some initial conditions for the simulation.

```
from modelbase.ode import mca

cccs, fccs = mca.get_response_coefficients_df(
    m,
    parameters=["Vm1", "Vm2", "Vm3"],
    y={"Glycolate": 0, "Glyoxylate": 0, "Glycine": 0},
)
```

Now we can for example plot both data frames as heatmaps

```
fig, axs = mca.plot_multiple(
    [cccs, fccs],
    titles=[
        "Concentration control coefficients",
        "Flux control coefficients",
    ],
    annotate=False,
)
```

which then gives us nice heatmaps shown in Fig. 5

7 Label Models

A common technique to measure concentrations of metabolites and fluxes *in vitro* is to use isotopes, e.g., ^{13}C, as labels which can be read out with little perturbation to the biological system. To simulate the dynamic distribution of isotopes, all possible labeling patterns for all intermediates need to be created. `modelbase` can automate creation of all 2^N isotope-specific versions of a chemical compound as well as mapping the respective isotopes to reactions provided an atom transition map. We will explore this using the

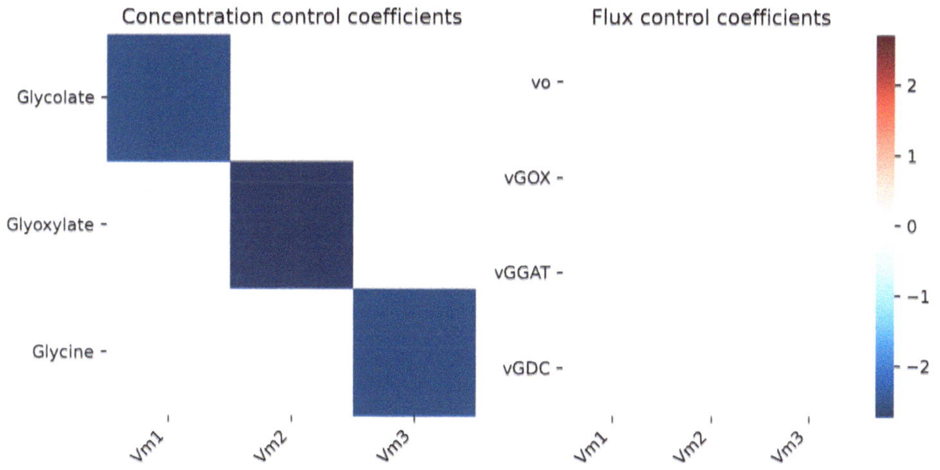

Fig. 5 Concentration control coefficients (left) and flux control coefficients (right) of the toy model

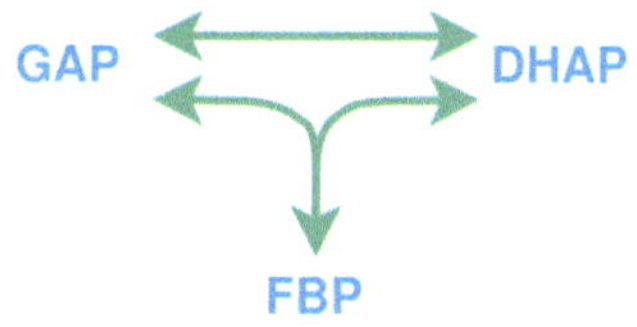

Fig. 6 Scheme of the toy label model. Note that the model is closed, in that there is no influx or outflux. *GAP* glyceraldehyde 3-phsophate, *DHAP* dihydroxyacetone phosphate, and *FBP* Fructose 1,6-bisphosphate

small, closed system in Fig. 6. The metabolites in the system, glyceraldehyde 3-phsophate (GAP), dihydroxyacetone phosphate (DHAP), and Fructose 1,6-bisphosphate (FBP), are upstream precursors of photorespiration and an integral part of the Calvin Benson Bassham cycle.

While you can build label models in `modelbase` from scratch, it is often a better idea to build a regular model first, to ensure valid behavior and then convert that model into a label model, which we will now do. The first step is to create a regular model of this system, where we separate the forward and backward direction of the reversible reactions. To illustrate why this is necessary, imagine an enzyme that rapidly converts between substrate and product in a way that there is just slightly more flux toward the product. In that case the overall rate of the reaction is very low, the label mixing between the substrate and product pool, however, is very fast.

```
def divide(x: float, y: float) -> float:
    return x / y


def mass_action_1(kf: float, s: float) -> float:
    return kf * s


def mass_action_2(kf: float, s1: float, s2: float) -> float:
    return kf * s1 * s2


m = Model(
    parameters={
        "kf_TPI": 1.0,
        "Keq_TPI": 21.0,
        "kf_Ald": 2000.0,
        "Keq_Ald": 7000.0,
    },
    compounds=["GAP", "DHAP", "FBP"],
)
m.add_derived_parameter("kr_TPI", divide, ["kf_TPI", "Keq_TPI"])
m.add_derived_parameter("kr_Ald", divide, ["kf_Ald", "Keq_Ald"])

m.add_reaction_from_args(
    rate_name="TPIf",
    function=mass_action_1,
    stoichiometry={"GAP": -1, "DHAP": 1},
    args=["kf_TPI", "GAP"],
)
m.add_reaction_from_args(
    rate_name="TPIr",
    function=mass_action_1,
    stoichiometry={"DHAP": -1, "GAP": 1},
    args=["kr_TPI", "DHAP"],
)
m.add_reaction_from_args(
    rate_name="ALDf",
    function=mass_action_2,
    stoichiometry={"DHAP": -1, "GAP": -1, "FBP": 1},
    args=["kf_Ald", "DHAP", "GAP"],
)
m.add_reaction_from_args(
    rate_name="ALDr",
    function=mass_action_1,
    stoichiometry={
        "FBP": -1,
        "DHAP": 1,
        "GAP": 1,
    },
    args=["kr_Ald", "FBP"],
)
```

Now that we have the regular model, we can build the label model from it using the `to_labelmodel` method, which we need to supply the number of label positions (e.g., carbons) per compound and the label transition maps.

```
lm = m.to_labelmodel(
    labelcompounds={"GAP": 3, "DHAP": 3, "FBP": 6},
    labelmaps={
        "TPIf": [2, 1, 0],
        "TPIr": [2, 1, 0],
        "ALDf": [0, 1, 2, 3, 4, 5],
        "ALDr": [0, 1, 2, 3, 4, 5],
    },
)
```

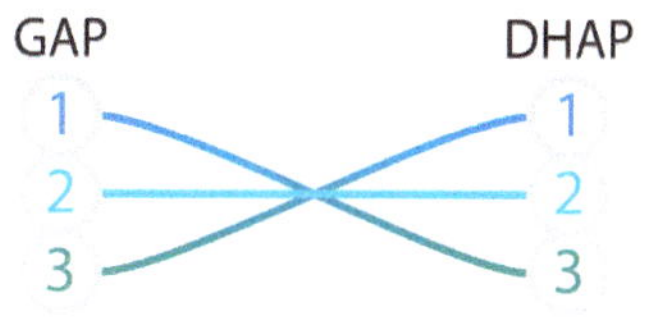

Fig. 7 Label transition map of triose phosphate isomerase (TPI)

Here is how you can understand how the label maps encode the transition of labels: for each label position of a substrate (in order) you specify the position of the product in which this label will end up. So the transition map of triose phosphate isomerase (TPI) reads `[2, 1, 0]`, which means that the first label of GAP ends up in the third position of DHAP, the second in the second and the third in the first, as illustrated in Fig. 7 (**Note 9**).

Now, just like we used the `Simulator` object before we can initialize and simulate our model, the only difference being that you need to have initial conditions for all of the isotopomers. You can use the `generate_y0` method to generate the initial conditions from concentration data of the non-labeled model, e.g., the steady-state concentrations from previous simulations. As a common case is that just a single metabolite or even just a single isotopomer is fully labeled the optional `label_positions` keyword argument can be used to specify which position that is. In the example below position 0 of GAP is fully labeled.

```
ls = Simulator(lm)
ls.initialise(lm.generate_y0(y_ss, label_positions={"GAP": 0}))
t, y = ls.simulate(20)
fig, ax = ls.plot_label_distribution_grid(
```

Now we can plot the label distributions of the respective metabolites. As this is quite a lot of information it makes sense to separate the label distributions of the different metabolites into multiple sub-plots, which we are going to do using the `plot_label_distribution_grid` method. As ^{13}C labels are usually denoted starting from 1 we will set the prefix of the legend to `"C"` and the initial index to `1`,

```
    ["GAP", "DHAP", "FBP"],
    ncols=3,
    sharey=False,
    relative=True,
    xlabels="Time [au]",
    prefix="C",
    initial_index=1,
)
```

which produces the plot in Fig. 8:

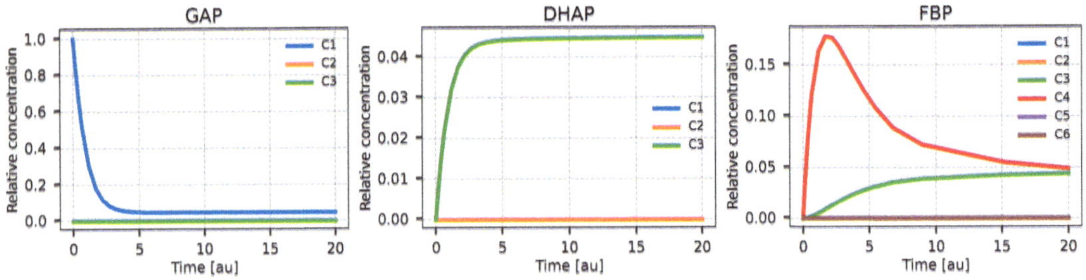

Fig. 8 Label distribution relative to the total amount of the respective metabolite for GAP, DHAP, and FBP **Note 10**

8 Notes

1. Python is one of many programming languages used in scientific computing. Python has the advantage of having a steep learning curve. Besides Python, the programming language Julia has become quite popular over the last couple of years.
2. The theory of differential equations and dynamic systems is a broad mathematical field with many applications in science and engineering. Besides ordinary differential equations (linear or nonlinear), there are partial, algebraic, or integro-differential equations. All of these can be applied to different biological problems. For models about metabolic pathways, ordinary or algebraic differential equations are often enough to make considerable progress and get a deeper understanding of the investigated system.
3. The choice of the Python distribution/package manager is arbitrary for our purposes in this chapter. However, we find that mamba is one of the fastest package management systems and is easy to learn. Therefore, it is preferable to other systems. It has many similarities with the conda management system in the anaconda distribution, which would be an alternative (https://www.anaconda.com, last accessed 05 December 2022).
4. We find that writing the system down and trying to understand it using mathematical techniques, such as linear stability analysis, is often beneficial before proceeding with the computer implementation. Even if the system is too complex for stability analysis, it is good advice to organize and write down the equation system before implementing it. This often can make the transfer from equations to computer code considerably easier.

5. There are a lot of openly available differential equation solvers. All of them are built using different methods. Consequently, the solvers considerably differ in the time used for solving a differential equation system. The choice of the differential equation solver depends on the size of the system and the underlying mathematics of the model implementation. We find that "cvode," used by `modelbase` from the sundials package, is an excellent solver suitable for many tasks.
6. The initial conditions should be compatible with constraints, for instance, the total pool size of metabolites.
7. By introducing a function describing a time-dependent external influx, we change the ordinary differential equation system from autonomous to non-autonomous. The mathematical treatment of non-autonomous differential equations needs different techniques than autonomous systems.
8. We find metabolic control analysis a nice technique to understand control in a biological network with solid mathematical base [13, 14]. MCA can be used to investigate metabolic systems with diverse behaviors. Note that metabolic control theory and engineering control theory, often applied in engineering sciences, are not the same. There have been efforts to connect both fields using a frequency domain approach in the past [15].
9. It is advisable to be aware of most label maps for the enzymes in metabolic pathways when building a label model. Also, writing the label maps down before proceeding to the computer implementation might be helpful.
10. Modelbase includes many methods to plot simulation outputs. However, if the user wants to customize their plots, it is possible to change almost any aspect using the package matplotlib [16]. The minimum for a plot is to label the axis meaningfully. modelbase reminds the user to label the plot if not done by the user.

Acknowledgements

This work was supported by the Deutsche Forschungsgemeinschaft (DFG), project ID 391465903/GRK 2466 (TN) and the EU's Horizon 2020 research and innovation programme under the Grant Agreement 862087 (MvA).

References

1. Saadat NP, Nies T, Van Aalst M, et al (2021) Computational analysis of alternative photosynthetic electron flows linked with oxidative stress. Front Plant Sci 12:750580. https://doi.org/10.3389/fpls.2021.750580
2. Stirbet A, Lazár D, Guo Y, Govindjee G (2020) Photosynthesis: basics, history and modelling. Anna Bot 126:511–537. https://doi.org/10.1093/aob/mcz171
3. Zhu X-G, Wang Y, Ort DR, Long SP (2013) e-photosynthesis: a comprehensive dynamic mechanistic model of C3 photosynthesis: from light capture to sucrose synthesis. Plant Cell Environ 36:1711–1727. https://doi.org/10.1111/pce.12025
4. Yokota A, Kitaoka S, Miura K, Wadano A (1985) Reactivity of glyoxylate with hydrogen perioxide and simulation of the glycolate pathway of C3 plants and Euglena. Planta 165:59–67. https://doi.org/10.1007/BF00392212
5. van Aalst M, Ebenhöh O, Matuszyńska A (2021) Constructing and analysing dynamic models with modelbase v1. 2.3: a software update. BMC Bioinform 22:1–15. https://doi.org/10.1186/s12859-021-04122-7
6. Ebenhöh O, van Aalst M, Saadat NP, et al (2018) Building mathematical models of biological systems with modelbase. J Open Res Softw 6. https://doi.org/10.5334/jors.236
7. Cleland WW (1963) The kinetics of enzyme-catalyzed reactions with two or more substrates or products: I. Nomenclature and rate equations. Biochimica et Biophysica Acta (BBA)-Specialized Section on Enzymological Subjects 67:104–137. https://doi.org/10.1016/0006-3002(63)91800-6
8. Cornish-Bowden A (2013) Fundamentals of enzyme kinetics. John Wiley & Sons, Weinheim
9. Sims PA (2009) An "Aufbau" approach to understanding how the King–Altman method of deriving rate equations for enzyme-catalyzed reactions works. J Chem Educ 86:385. https://doi.org/10.1021/ed086p385
10. Meurer A, Smith CP, Paprocki M, et al (2017) SymPy: symbolic computing in Python. PeerJ Comput Sci 3:e103. https://doi.org/10.7717/peerj-cs.103
11. Kacser H (1973) The control of flux. In: Symposia of the Society for Experimental Biology, pp 65–104
12. Heinrich R, Rapoport TA (1974) A linear steady-state treatment of enzymatic chains: general properties, control and effector strength. Eur J Biochem 42:89–95. https://doi.org/10.1111/j.1432-1033.1974.tb03318.x
13. Heinrich R, Schuster S (1996) Metabolic control analysis. In: The regulation of cellular systems. Springer, London, pp 138–291
14. Fell D, Cornish-Bowden A (1997) Understanding the control of metabolism. Portland Press London, London
15. Ingalls BP (2004) A frequency domain approach to sensitivity analysis of biochemical networks. J Phys Chem B 108:1143–1152. https://doi.org/10.1021/jp036567u
16. Hunter JD (2007) Matplotlib: a 2D graphics environment. Comput Sci Eng 9:90. https://doi.org/10.1109/MCSE.2007.55

Chapter 19

Using Publicly Available RNA-seq Data for Expression Analysis of Genes of Interest

Donat Wulf, Andrea Bräutigam, and Marion Eisenhut

Abstract

RNA-seq data in publicly available repositories enable the efficient reanalysis of transcript abundances in existing experiments. Graphical user interfaces usually only allow the visual inspection of a single gene and of predefined experiments. Here, we describe how experiments are selected from the Sequence Read Archive or the European Nucleotide Archive, how data is efficiently mapped onto a reference transcriptome, and how global transcript abundances and patterns are inspected. We exemplarily apply this analysis pipeline to study the expression of photorespiration-related genes in photosynthetic organisms, such as cyanobacteria, and to identify conditions under which photorespiratory transcript abundances are enhanced.

Key words Illumina sequencing, Transcriptomics, RNA-seq, Reference genome, Kallisto, Quantification, Coefficient of variation

1 Introduction

RNA-seq analysis using Illumina sequencing has become common practice within the last two decades. In consequence, the sequence read archives (SRA) at the National Center for Biotechnology Information (NCBI [1]) and European Nucleotide Archive (ENA [2]) have become filled with published and unpublished experiments deposited as FAIR (Findable, Accessible, Interoperable, Reusable data) data [3], which can be reanalyzed with the correct tools. These data allow the study of genes of interest, such as photorespiratory gene expression in various conditions and in various mutants without additional wet-lab experiments. Currently, some of these experiments in some organisms are accessible via web interface, i.e., the eFP genome browser [4], which is available for many plant model organisms and for curated datasets. Using this data, scientists inspect transcript abundance patterns and generate hypotheses about their gene or genes of interest [4]. However,

Berkley J. Walker (ed.), *Photorespiration: Methods and Protocols*, Methods in Molecular Biology, vol. 2792,
https://doi.org/10.1007/978-1-0716-3802-6_19,

this inspection is limited to the curated datasets available in graphical user interfaces.

To inspect changes in transcript abundances for any experiment of interest and any organism in the databases, the data needs to be downloaded, mapped onto a reference genome, and converted into a human-readable format. In this method, we will be exploring the photorespiratory pathway in the model cyanobacterium *Synechocystis* sp. PCC 6803. The protocol can be also applied for the analysis of alga or plant data. In photosynthetic eukaryotes, the sequencing library is usually generated by poly-A enrichment [5]. For bacteria, RNA-seq reads in the databases were generated from isolated bacterial RNA, usually depleted of rRNA, which were converted into a sequencing library [6–8]. In this library, original RNA molecules are represented by DNA fragments that are flanked by adapters with sequencing information and with bar codes. For quantification, the library is sequenced on an Illumina instrument [9] either in single-end mode, meaning the sequence fragment is read from one side only, or in paired-end mode, meaning the sequence fragment is read from both ends with an unsequenced gap in between. The length of sequencing varies but is generally long enough to obtain a convincing match during mapping. The sequencing machine outputs sequencing reads, which are already mostly devoid of sequencing adaptors and bar codes. They can directly be mapped onto references.

For mapping the RNA-seq data onto a reference sequence, multiple methods exist [10]. If only fast quantification is the goal, kallisto is a quick, convenient, and efficient tool to gather the estimated counts of each transcript in the read library and to gather normalized expression values per transcript output as transcripts per million transcripts (TPM) [10, 11].

Successful analysis of publicly available data requires six steps: (1) setting up a computer environment with the necessary programs, (2) choosing the data from an SRA, (3) the transcriptome to map the data on, (4) running the download, (5) mapping pipeline, and (6) visualizing the data. Here, we describe the analysis pipeline using the model cyanobacterium *Synechocystis* sp. PCC 6803 as an example of a photosynthetic organism. However, the pipeline is completely transferable to algae or plants.

2 Materials

1. A Linux computing environment, such as Ubuntu, Fedora, or similar; alternatively, Apple computer system (*see* **Note 1**).
2. Repository at GitHub for the download and mapping pipeline: https://github.com/DonatWulf/sra_mapping_script_0.2

3 Methods

3.1 Setting Up Computer Environment

1. Install kallisto by typing:
   - `git clone` https://github.com/pachterlab/kallisto.git
   - `cd kallisto`
   - `mkdir build`
   - `cd build`
   - `cmake ..`
   - `make`
   - `make install`

   (*see* **Notes 2** and **3**)
2. Install SRA-Toolkit:

   Follow the **steps 1** to **3** (downloading, installing, and configuration) in this manual: https://github.com/ncbi/sra-tools/wiki/HowTo:-fasterq-dump
3. Install R by typing:
   - `sudo apt update`
   - `sudo apt install r-base`
   - `sudo apt install r-base-dev`
   - `sudo apt install libcurl4-openssl-dev libssl-dev libxml2-dev`
4. Open R and install the packages tidyverse, doParallel, jsonlite, and argparser by typing
   - `R`
   - `install.packages(c("tidyverse", "doParallel","jsonlite","argparser")`

3.2 Choosing the Data to Be Mapped

1. Open https://www.ncbi.nlm.nih.gov/sra/advanced to access the advanced search.
2. Use the builder to limit the inquiry to "Organism" and choose "synechocystis pcc6803"; the index list can be helpful for choosing. Note how the search query on the top of the page builds ("synechocystis pcc6803"[Organism]). Use the builder to limit the inquiry to "Platform" and choose "illumina" and the "source" to "transcriptomic." The final search query should now read (("synechocystis pcc6803"[Organism]) AND "illumina"[Platform]) AND "transcriptomic"[Source] and be executed (*see* **Notes 4** and **5**).
3. To inspect and decide on the files, use the "Send to:" button to send the summary to a file from which the experiments of interest are chosen. Also use "Send to:" button to send the RunInfo to a file.

4. From the resulting file, select the RNA-seq experiments of interest. You will need the RunIDs and the RunInfo later during mapping, so archive these files.

3.3 Choosing the Reference Sequence

1. Search for *Synechocystis* sp. PCC 6803 genomes at a repository, such as NCBI, and decide on the genome to use. We recommend the original genome publication [12, 13] (*see* **Note 6**) of which the data can currently be accessed via https://ftp.ncbi.nlm.nih.gov/genomes/all/GCF/000/009/725/GCF_000009725.1_ASM972v1/.

   If you want to work with algal or plant data, it is crucial to use the correct reference for the studied species (*see* **Note 7**).
2. Rename the transcript file "your_transcriptome.fasta" (*see* **Note 8**).

3.4 Running the Download

For mapping, download the reference transcriptome sequence. The feature_table.txt file in the same ftp folder would be useful to link the gene IDs from the coding sequence file to other gene identification systems.

3.5 Running the Mapping Pipeline

1. Download the scripts deposited at GitHub: https://github.com/DonatWulf/sra_mapping_script_0.2
2. Follow the readme for up-to-date instructions on how to use the pipeline.
3. Edit the config file with your installation paths noted during program installation. Examples are provided to test if everything is correctly installed.
4. To check if all tools and packages are installed, run the example on the command line from the location of the GitHub script:
   - `Rscript --vanilla multiSRAmapper.R -s 'examples/SRA_minimal_example.txt' -t 'examples/Solyc_example_transcriptom.fasta'`
5. To inform yourself about the possible options, run the help command:
   - `Rscript --vanilla multiSRAmapper.R --h`

You will find different command options:

--c, --check_reads: if set the number of processed reads is compared to the provided number of reads in the input files

-c, --combine: use this flag to just combine individual mappings into the expression matrix

-t, -transcriptome: transcriptome fasta file to generate the kallisto index and map the reads

-r, -retries: number of retries for failed downloads [default: 3]

-p, -paralell_downloads: how many downloads are run in parallel. This increases the storage requirement. At some point, you get diminishing returns [default: 2]

-k, -paralell_kallisto: how many threads kallisto should use for mapping. [default: number of available threads]

-s, -SRA_info: SRA info file containing the RunID, ExperimentID, LibraryLayout, Stranded, the number of spots comma separated. If you want to map each run individually, you need to assign an individual ExperimentID or use the RunID as the ExperimentID

-sd, -standard_deviation: standard deviation used for mapping [default: 20]

-fl, -fragment_length: fragment length used for mapping [default: 200]

-o, -output_file: output file name prefix. [default: expression_matrix]

6. From the chosen experiments (*see* Subheading 3.2), a tab-delimited text file with the ExperimentIDs and corresponding RunIDs, strandedness (*see* **Note 9**), and the number of spots and sequencing mode needs to be compiled. Use the example file provided as "examples/SRA_minimal_example.txt" as the guide for formatting the file. The Run IDs, Experiment IDs number of spots and sequencing mode are information provided by the SRA-RunInfo file (*see* **step** 4 in Subheading 3.2, *see* **Note 10**). Name the file "your_SRAinfo_file.txt"
7. Adjust the number of simultaneous downloads to fit within your available disk space. The pipeline does not rerun already mapped experiments (*see* **Note 11**).
8. Run the script with your own data set:
   - `Rscript --vanilla multiSRAmapper.R -s 'your_SRAinfo_file.txt' -t 'your_transcriptome.fasta'`
9. If multiple downloads are to be combined run:
   - `Rscript --vanilla multiSRAmapper.R --combine.`

   This command reads all the expression files of the experiments and combines them into the expression matrix. The expression matrix contains the genes in rows and the RunIDs in columns.
10. It is often useful to generate at least one statistical parameter to check the experiment. Summing the number of estimated counts in each column (i.e., via the =sum() function in Excel or an apply function in R, apply(dfr, 2, sum)) will show how

many reads could be mapped in each experiment. More reads mapped represent a higher depth of sequence coverage.

3.6 Visualizing the Data

1. Excel or R can be used to extract information about the genes of interest. Using Excel, generate the list of photorespiratory genes curated from databases, such as KEGG [14], AraCyc [15], or genes of interest in a second sheet.
2. Given a larger table of mapped experiments (i.e., [16]), one can determine the coefficient of variation (CoV) for all genes and compare it against the photorespiratory genes. For each gene, Excel can generate the mean (=AVERAGE()) and the standard deviation (=ST.DEV.P) of all experiments from which the CoV is calculated. The CoV of photorespiratory genes is lower than the median CoV except for phosphoglycolate phosphatase (*pglp*) genes, when compared against the backdrop of all genes and compared with house-keeping (e.g., *petB*) and variable genes (e.g., *flv2, flv4*) (*see* Fig. 1a, *see* **Note 12**).
3. Abundance of information from the expression matrix (*see* **Note 13**) can be transferred to lists of genes of interest in Excel using the =VLOOKUP() function using the gene ID as the key, followed by the column index of the column to be transferred and the "false" moniker to force exact matches. In R, the merge() or left_join() will accomplish merging of data frames. This method will generate a smaller table with only photorespiratory data that can be visualized (*see* Fig. 1b).

4 Notes

1. Set-up and permissions may be different on an Apple computer.
2. Alternatively, follow this manual: https://github.com/pachterlab/kallisto/blob/master/INSTALL.md
3. Please note that if start and end sites of transcripts or operon structures are the analytical goal, or small or noncoding RNAs, methods other than kallisto, which employ direct read alignment, are needed [17].
4. The methods outlined use FAIR data and hence are prone to FAIR data pitfalls. Only if the data to be reanalyzed is deposited fairly in the databases with sufficient and correct annotation, reanalysis is possible. In case, a reanalysis fails, it is paramount to check the mapping parameters used in kallisto, as incorrect mapping parameters lead to skewed or unreadable data outputs. Sometimes data for a published RNA-seq experiment is unavailable in the databases since it was not deposited. In that case, contact with the authors should be made to obtain the

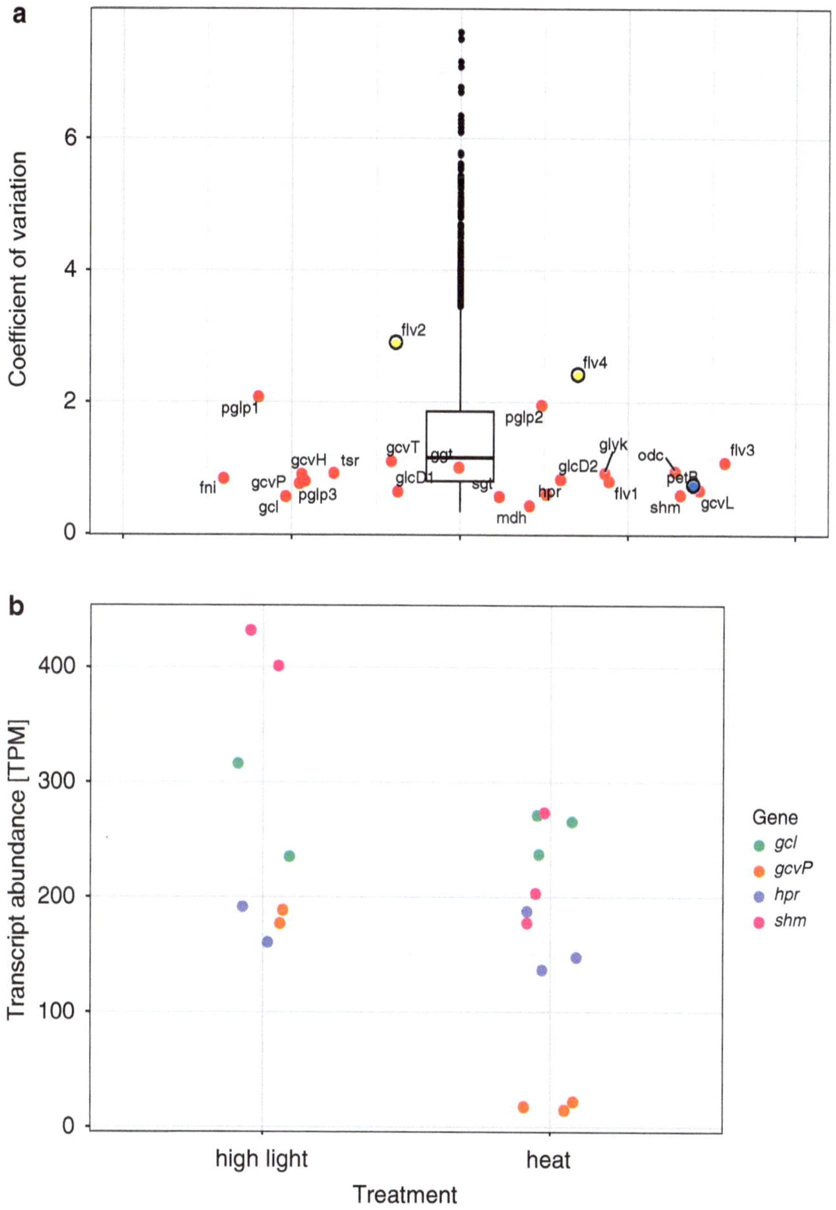

Fig. 1 Visualization of gene expression data. (**a**) Coefficient of variation for photorespiratory genes in *Synechocystis*. In comparison to the photorespiratory genes (according to [24], depicted as red points), *pglp1* (*slr*0458), *pglp2* (*sll*1349), *pglp3* (*slr*0586), *glcD1* (*sll*0404), *glcD2* (*slr*0806), *ggt* (*slr*0006), *gcvH* (*slr*0879), *gcvL* (*slr*1096), *gcvP* (*slr*0293), *gcvT* (*sll*0171), *shm* (*sll*1931), *sgt* (*sll*1559), *hpr* (*sll*1908), *gcl* (*sll*1981), *tsr* (*slr*0229), *odc* (*sll*1358), genes for flavodiiron proteins *flv2* (*sll*0219) and *flv4* (*sll*0217) are variable in expression (depicted as yellow points), while *petB* (*slr*0342) is a representative for a house-keeping gene [25] (depicted as blue point). (**b**) Transcript abundances of selected photorespiratory genes (*gcl*, *gcvP*, *hpr*, *shm*) after high light (SRX13465627 and SRX13465628) or heat treatment (SRX13465630, SRX13465631, and SRX13465632)

raw data files. If data is to be further analyzed for statistically significant differences (not covered in these methods), it is recommended to check the distribution of replicates in a principal component analysis prior to continuing to statistical analyses.

5. Care should be taken to avoid mapping small RNAs, which will not represent mRNA abundances or specialized depletion or enrichment experiments. It can be useful to check the publications associated with read depositions. Please note that the annotations in the file have been entered by other users and may or may not be accurate. The archives only check for formal errors.
6. Multiple sequenced genomes for a species and their corresponding derived transcriptome sequences are also deposited in the archives. For the model cyanobacterium *Synechocystis* sp. PCC6803, the NCBI archive contains 11 genome sequences (data retrieved 02/2023) of which GCA_000009725.1 represents the original sequenced genome and plasmids [12, 13].
7. Reliable resources for plants are ensemblPlants [18], JGI Phytozome [19], or NCBI [20]. It is good practice to check, which reference is used in current publications. You are looking for the transcriptome or cDNA. The nomenclature is different from one resource to the other.
8. In case the ftp link moves, follow a link trail through NCBI after searching for the genomes.
9. The strandedness needs to be determined by reading the methods of corresponding paper or with the tool "how_are_we_stranded_here": https://github.com/signalbash/how_are_we_stranded_here.

   Bacterial RNA-seq usually occurs in stranded mode [6, 21].
10. While the sequence read archive should provide information on the sequence mode, which then is translated into the settings for kallisto, the information might be incorrect necessitating test runs for mapping each experiment.
11. Make sure that enough disk space is available to store the fastq files until they are mapped. If the download of some experiments terminates, it is advised to restart the pipeline. In rare cases, a download or the mapping of some datasets is not possible.
12. CoVs were calculated for 204 different experiments. A low CoV does not necessarily mean that a gene is expressed similarly in all conditions, as evidenced by *gcvP* and *shm* (*see* Fig. 1).

13. Estimated counts can be further processed to assess differential expression, for example, via edgeR [22] or DEseq2 [23], if desired.

Acknowledgments

This work was supported by the Deutsche Forschungsgemeinschaft (DFG) through the grant EI 945/3-2 and SFB1535 – Project ID 458090666 to M.E. Core funding by Bielefeld University to A.B. is acknowledged.

References

1. NCBI Sequence Read Archive. https://www.ncbi.nlm.nih.gov/sra. Accessed 7 Mar 2023
2. ENA European Nucleotide Archive. https://www.ebi.ac.uk/ena/browser/home. Accessed 7 Mar 2023
3. Wilkinson MD, Dumontier M, Aalbersberg IJJ et al (2016) The FAIR guiding principles for scientific data management and stewardship. Sci Data 3:160018. https://doi.org/10.1038/sdata.2016.18
4. Winter D, Vinegar B, Nahal H et al (2007) An "Electronic fluorescent pictograph" browser for exploring and analyzing large-scale biological data sets. PLoS One 2:e718. https://doi.org/10.1371/journal.pone.0000718
5. Zhao W, He X, Hoadley KA et al (2014) Comparison of RNA-Seq by poly (A) capture, ribosomal RNA depletion, and DNA microarray for expression profiling. BMC Genomics 15:419. https://doi.org/10.1186/1471-2164-15-419
6. Anfelt J, Hallström B, Nielsen J et al (2013) Using transcriptomics to improve butanol tolerance of *Synechocystis* sp. strain PCC 6803. Appl Environ Microbiol 79:7419–7427. https://doi.org/10.1128/AEM.02694-13
7. Qiao J, Huang S, Te R et al (2013) Integrated proteomic and transcriptomic analysis reveals novel genes and regulatory mechanisms involved in salt stress responses in *Synechocystis* sp. PCC 6803. Appl Microbiol Biotechnol 97:8253–8264. https://doi.org/10.1007/s00253-013-5139-8
8. Huang S, Chen L, Te R et al (2013) Complementary iTRAQ proteomics and RNA-seq transcriptomics reveal multiple levels of regulation in response to nitrogen starvation in *Synechocystis* sp. PCC 6803. Mol Biosyst 9:2565. https://doi.org/10.1039/c3mb70188c
9. Goodwin S, McPherson JD, McCombie WR (2016) Coming of age: ten years of next-generation sequencing technologies. Nat Rev Genet 17:333–351. https://doi.org/10.1038/nrg.2016.49
10. Yi L, Liu L, Melsted P, Pachter L (2018) A direct comparison of genome alignment and transcriptome pseudoalignment. bioRxiv:444620. https://doi.org/10.1101/444620
11. Bray NL, Pimentel H, Melsted P, Pachter L (2016) Near-optimal probabilistic RNA-seq quantification. Nat Biotechnol 34:525–527. https://doi.org/10.1038/nbt.3519
12. Kaneko T, Sato S, Kotani H et al (1996) Sequence analysis of the genome of the unicellular cyanobacterium *Synechocystis* sp. strain PCC6803. II. Sequence determination of the entire genome and assignment of potential protein-coding regions. DNA Res 3:109–136. https://doi.org/10.1093/dnares/3.3.109
13. Kaneko T (2003) Structural analysis of four large plasmids harboring in a unicellular cyanobacterium, *Synechocystis* sp. PCC 6803. DNA Res 10:221–228. https://doi.org/10.1093/dnares/10.5.221
14. Kanehisa M, Sato Y, Kawashima M et al (2016) KEGG as a reference resource for gene and protein annotation. Nucleic Acids Res 44:D457–D462. https://doi.org/10.1093/nar/gkv1070
15. Zhang P, Foerster H, Tissier CP et al (2005) MetaCyc and AraCyc. Metabolic pathway databases for plant research. Plant Physiol 138:27–37. https://doi.org/10.1104/pp.105.060376
16. Reis M, Brandenburg F, Knopp M et al (2023) Hemi Manganese Exporters 1 and 2 enable manganese transport at the plasma membrane in cyanobacteria.

bioRxiv:2023.02.16.528846. https://doi.org/10.1101/2023.02.16.528846

17. Brenes-Álvarez M, Olmedo-Verd E, Vioque A, Muro-Pastor AM (2016) Identification of conserved and potentially regulatory small RNAs in heterocystous cyanobacteria. Front Microbiol 7. https://doi.org/10.3389/fmicb.2016.00048
18. EnsemblPlants. https://plants.ensembl.org. Accessed 7 Mar 2023
19. JGI Phytozome. https://phytozome-next.jgi.doe.gov. Accessed 7 Mar 2023
20. NCBI. https://www.ncbi.nlm.nih.gov. Accessed 7 Mar 2023
21. Mitschke J, Georg J, Scholz I et al (2011) An experimentally anchored map of transcriptional start sites in the model cyanobacterium *Synechocystis* sp. PCC6803. Proc Natl Acad Sci 108: 2124–2129. https://doi.org/10.1073/pnas.1015154108
22. Robinson MD, McCarthy DJ, Smyth GK (2010) <tt>edgeR</tt>: a bioconductor package for differential expression analysis of digital gene expression data. Bioinformatics 26:139–140. https://doi.org/10.1093/bioinformatics/btp616
23. Love MI, Huber W, Anders S (2014) Moderated estimation of fold change and dispersion for RNA-seq data with DESeq2. Genome Biol 15:550. https://doi.org/10.1186/s13059-014-0550-8
24. Eisenhut M, Ruth W, Haimovich M et al (2008) The photorespiratory glycolate metabolism is essential for cyanobacteria and might have been conveyed endosymbiontically to plants. Proc Natl Acad Sci U S A 105:17199–17204. https://doi.org/10.1073/pnas.0807043105
25. Pinto F, Pacheco CC, Ferreira D et al (2012) Selection of suitable reference genes for RT-qPCR analyses in cyanobacteria. PLoS One 7:e34983. https://doi.org/10.1371/journal.pone.0034983

Chapter 20

Identification and Functional Characterization of *cis*-Regulatory Elements of Key Photorespiratory Genes in Response to Short-Term Abiotic Stress Conditions

Debarati Basu, Casey Butler, Mary Beth Rollins, and Paul South

Abstract

The *cis*-regulatory elements (CREs) are the short stretches of noncoding DNA upstream of a gene, which play a critical role in fine-tuning gene expression. Photorespiration is a multi-organellar, energy-expensive biochemical process that remains intricately linked to photosynthesis and is conserved in plants. Recently, much focus has been devoted in generating plants with engineered alternative photorespiratory bypasses to enhance photosynthetic efficiency without compromising the beneficial aspect of photorespiration. Varied constitutive or inducible promoters for generating transgenic plants harboring multiple transgenes have been introduced over years; however, most of them suffer from unintended effects. Consequently, a demand for synthetic tunable promoters based on canonical CRE signatures derived from native genes is on the rise. Here, in this chapter, we have provided a detailed method for in silico identification and characterization of CREs associated with photorespiration. In addition to the detailed protocol, we have presented an example of a typical result and explained the significance of the result. Specifically, the method covers how to identify and generate tunable synthetic promoters based on native CREs using three key photorespiratory genes from *Arabidopsis* and two web-based tools, namely, PlantPAN3.0 and AthaMap. Finally, we have also furnished a protocol on how to test the efficacies of the synthetic promoters harboring predicted CREs using transient tobacco expression coupled with luciferase-based promoter assay in response to ambient conditions and under short-term abiotic stress conditions.

Key words *cis*-regulatory elements, Photorespiration, Abiotic stress, Arabidopsis

1 Introduction

Photorespiration (PR) is an inevitable metabolic pathway shared among all oxygenic photoautotrophs. Plants utilize PR to convert 2-phosphoglycolate (2PG), an inhibitory by-product of Rubisco's oxygenase activity, into 3-phosphoglycerate (3PGA), an intermediate of the Calvin–Benson cycle [1]. PR results in a loss of fixed CO_2, energy, and reducing equivalents, thereby severely hindering crop productivity [2]. Considering the negative impact of PR on crop productivity, substantial effort has been made to characterize the

Berkley J. Walker (ed.), *Photorespiration: Methods and Protocols*, Methods in Molecular Biology, vol. 2792,
https://doi.org/10.1007/978-1-0716-3802-6_20,

genes involved in the PR pathway [3–5]. While PR is a wasteful metabolic process in the sense that it requires energy and loses previously fixed carbon, it is also a critical recycling pathway for 2PG and has also been proposed to play an important role in nitrogen and sulfur metabolism and in establishing redox homeostasis [6–9]. Although considerable effort has been devoted on characterizing PR genes, little is known about how these genes are regulated. Such knowledge on PR gene regulation can be instrumental in designing an alternative PR pathway to minimize PR loss without affecting its native axillary functions. Furthermore, the rate of PR drastically increases in response to elevated temperature, high light intensity, and low CO_2 [10–13].

PR requires the cooperation of three different organelles: the chloroplast, the mitochondrion, and the peroxisome plus the cytoplasm. The first step in the PR pathway is the dephosphorylation of 2-PG by 2-phosphoglycolate phosphatase (PGLP) to produce glycolate [14]. Also, two plastidal membrane-transporters PLGG1 (glycolate/glycerate transporter 1) and BASS6 (bile acid sodium symporter) are critical for transporting glycolate from chloroplasts [15, 16]. We selected these three genes for identifying common stress-responsive *cis*-regulatory (CREs).

A fundamental question in biology is how complex patterns of a gene are precisely regulated. Synthetic promoters based on native CREs are ideal as such promoters can be designed to be scalable, tunable, and more consistent than the available inducible and/or constitutive promoters (*see* **Note 1**). A recent report has suggested that common stress-responsive motifs play a critical role for *AtPLGG1*, *AtBASS6*, and *AtPGLP* in mediating response to elevated temperature and high-intensity light stress [17]. Here, we present a detailed method covering the identification of putative stress-responsive CREs common among three core PR genes to characterization of the synthetic promoters under ambient and short-term abiotic stress conditions (Fig. 1).

2 Materials

2.1 Plant/Bacterial

1. *Nicotiana benthamiana* seeds.
2. *Escherichia coli* DH10β electro competent cells.
3. *Agrobacterium tumefaciens* (GV3101).
4. The *p19* plasmid is from tomato bushy stunt virus for suppression of endogenous silencing.
5. Nano-Luciferase coding region.

Identification of putative stress-responsive *cis*-regulatory elements (CRE) using three core photorespiratory genes (PR)

↓

Select 2-kb upstream regions from predicted start codon (ATG) of three PR genes (*AtPGLP*, *AtPLGG1*, *AtBASS6*)

↓

Submit the sequences to two independent web-based promoter analysis tools, AthaMap1 and PlantPAN 3.0

↓

Select common CREs involved in abiotic stress with 80-90 % score

↓

Synthesize the seven selected CREs and fuse with minimal promoter and 1X35SCE

↓

Clone the CREs into transcriptional unit using GoldenGate cloning strategy

↓

Transiently express the synthetic promoter fused to luciferase reporter in tobacco leaves

↓

Subject the agroinfiltrated tobacco plants to stress treatment and perform *in-vitro* luciferase assay

Fig. 1 Flow chart depicting the identification and functional characterization of synthetic promoter containing native *cis*-regulatory elements shared by three core photorespiratory genes

2.2 Plant/Bacterial Growth

1. Sun Gro potting mix.
2. Individual 4′ × 3.5′ pots.
3. Soda lime, Litholyme Carbon Dioxide Absorbent.
4. Bacterial media—Luria-Bertani; for 1 L weigh 10 g of Tryptone, 10 g of Sodium Chloride, and 5 g of Yeast extract. Dissolve all the chemicals in 1 L of water. For making solid LB media, add 10 g of Bacto agar before autoclaving. Autoclave the media at 121 °C for 30 min at 15 psi. Cool down the media at 60 °C and pour ~25 mL per 100-mm plate

supplemented with appropriate antibiotics, including Spectinomycin, Carbenicillin, Kanamycin, or Gentamycin. For selection of L1 Golden Gate constructs, spread X-gal over the appropriate plates for selection of the transformed colonies.

5. DNase/RNase-free water.

2.3 Buffers

1. Agrobacterium infiltration buffer—Add 10 mM MES, 10 mM $MgCl_2$, and 150 μM acetosyringone). Dissolve all the chemicals in water and set the pH at 5.8. Filter sterilize the buffer using filtering units fitted with 0.22 μm filter and a pump. Store the buffer at 4 °C refrigerator.
2. Total protein extraction buffer—Add 20 mM Tris–HCl, 150 mM NaCl, and 5 mM $MgCl_2$. Dissolve all the chemicals in water and set the pH at pH 7.5. Store the buffer at 4 °C refrigerator. Add 5 mM DTT just before use.

2.4 Equipment

1. Percival growth chamber.
2. Light meter.
3. Autoclave.
4. PCR machine.
5. Plate reader.
6. Electroporator.
7. Incubator shaker.
8. Nanodrop, Take 3 Microvolume plate.
9. Bench-top centrifuge.
10. Spectrophotometer.

2.5 Consumables

1. Sterile RNAse/DNAse-free filter tips.
2. Sterile microcentrifuge tubes.
3. 1ml syringes without needles.
4. Clear 96-well plates.
5. Opaque 96-well plates.
6. Electro-cuvettes.
7. 15 mL disposable culture tubes.
8. Sterile petri-plates.
9. PCR tubes.
10. DNA Polymerase.
11. Dual Luciferase Reporter Assay System kit containing luciferase assay substrate, luciferase assay buffer, stop and glow buffer, and stop and glow substrate.

2.6 Web-Based Tools/Database/ Software

1. AthaMap (Hehl et al., 2016) (http://www.athamap.de/).
2. PlantPAN 3.02 (Chow et al., 2019) (http://plantpan.itps.ncku.edu.tw/).
3. TAIR (https://www.arabidopsis.org/index.jsp).
4. GraphPad Prism version 7.04.

3 Methods

3.1 Plant Growth Conditions

1. Plant *Nicotiana benthamiana* seeds in Sun Gro potting mix and then grow in a Percival growth chamber with light intensity set for 150 μmol photons $m^{-2}\ s^{-1}$. Around 24 plants are grown in individual pots.
2. Maintain the growth chambers at 12 h day/night cycle photoperiod with day temperatures maintained at 22 °C and night temperatures reaching 20 °C. Maintain all plants under relative humidity of 50%.
3. For ambient growth conditions, grow plants at atmospheric CO_2 pressure of about 400 ppm, light intensity of 150 μmol photons $m^{-2}\ s^{-1}$, and temperatures of 22 °C (*see* **Note 2**).
4. To implement high light intensity stress, exposed plants to 1200 μmol photons $m^{-2}\ s^{-1}$ and temperature regimes of 25 °C.
5. To achieve elevated temperatures grow plants at 32 °C under 150 μmol photons $m^{-2}\ s^{-1}$ of light intensity.
6. For low CO_2 expose plants to 200 ppm of CO_2.

3.2 *In Silico* Analysis of Promotor Regions

1. Select genes for CRE identification—As an example, we selected three Arabidopsis photorespiratory genes including *AtPGLP* (2-phosphoglycolate phosphatase), *AtBASS6* (Bile Acid Sodium Symporter 6), *AtPLGG1* (plastidic glycolate glycerate transporter) are selected for CRE identification.
2. Using the seq-viewer browser hosted by TAIR, select 2-kb upstream regions of the predicted start codon for all three genes.
3. Use "co-occurrence" tool under "Gene group analysis" function of PlantPAN 3.0 for identifying shared CREs (*see* **Note 3**).
4. For analysis using AthaMap, use "Gene Analysis" function by setting the upstream region at −2000 bp and "% Restriction to highly conserved TF binding sites" at 80.

3.3 Cloning Desired Constructs

Clone and assemble desired CRE plasmids using the Golden Gate modular cloning system containing standardized overhangs [18, 19] (*see* **Notes 4**, **5**, **6**, and **7**).

1. To test the function of putative stress-responsive CREs, select four copies of individual CREs along with a single copy of 35SCRE (CaMV35S promoter) followed by a TATA motif fused to a luciferase reporter gene. Introduce minimal promoter (MP), which contains no activation elements upstream of the TATA motif as a negative control, whereas full length p35SFL (CaMV35S promoter) is used as a positive control for luciferase activity. For a thorough comparison of the strength of the selected CREs, include a multiple CRE (MCRM) comprising all the selected putative CREs into one construct (Fig. 2).
2. Synthesize the Level 0 (L0) genes either linear fragments or circular plasmids. These synthesized gene fragments or circular plasmids contain compatible overhangs that are defined by the Golden Gate System [18, 19] (Fig. 3a).

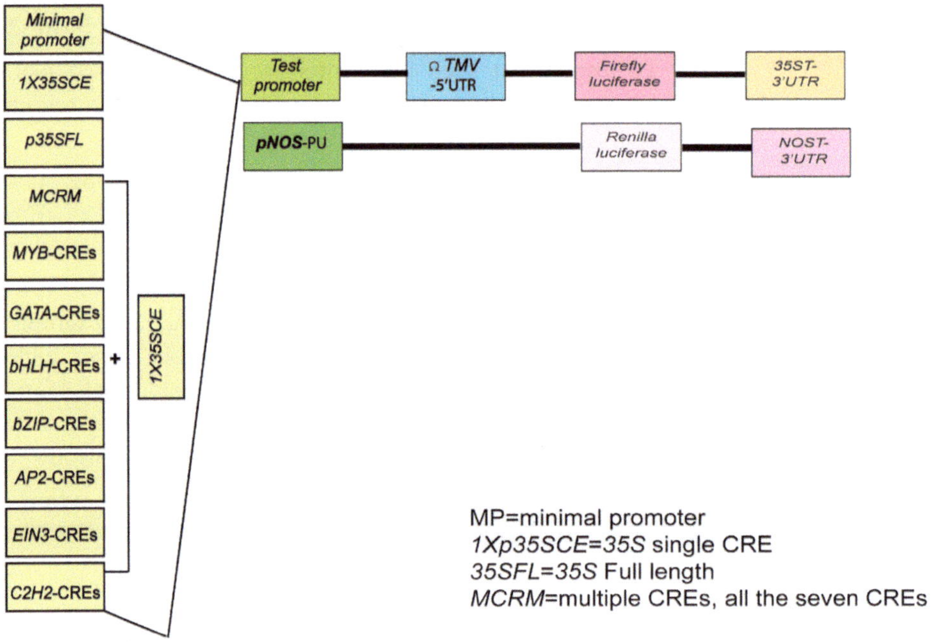

Fig. 2 Details of the transcriptional units containing seven putative stress-responsive CREs shared by the three photorespiratory genes. Seven different CREs were identified by submitting predicted promoter regions of *AtPGLP*, *AtPLGG*, and *AtBASS6* genes and using AthaMap and PlantPAN 3.0. As shown in the illustration, individual CREs are linked with short nonfunctional linker sequences along with proximally placed *1X35SCRE* promoter for luciferase-based promoter-efficiency assay. Acronyms explained as follows- *ΩTMV*, *5′UTR* from tobacco mosaic virus; MP, minimal promoter devoid of any CREs; *1Xp35SCE*, single CRE of CaMV; *pCaMV35SFL* or *p35SFL*, full length CaMV promoter, *pNOS-PU*, Agrobacterium Nopaline Synthase promoter and 5′UTR; *35ST*, CaMV terminator; *NOST*, Agrobacterium Nopaline Synthase

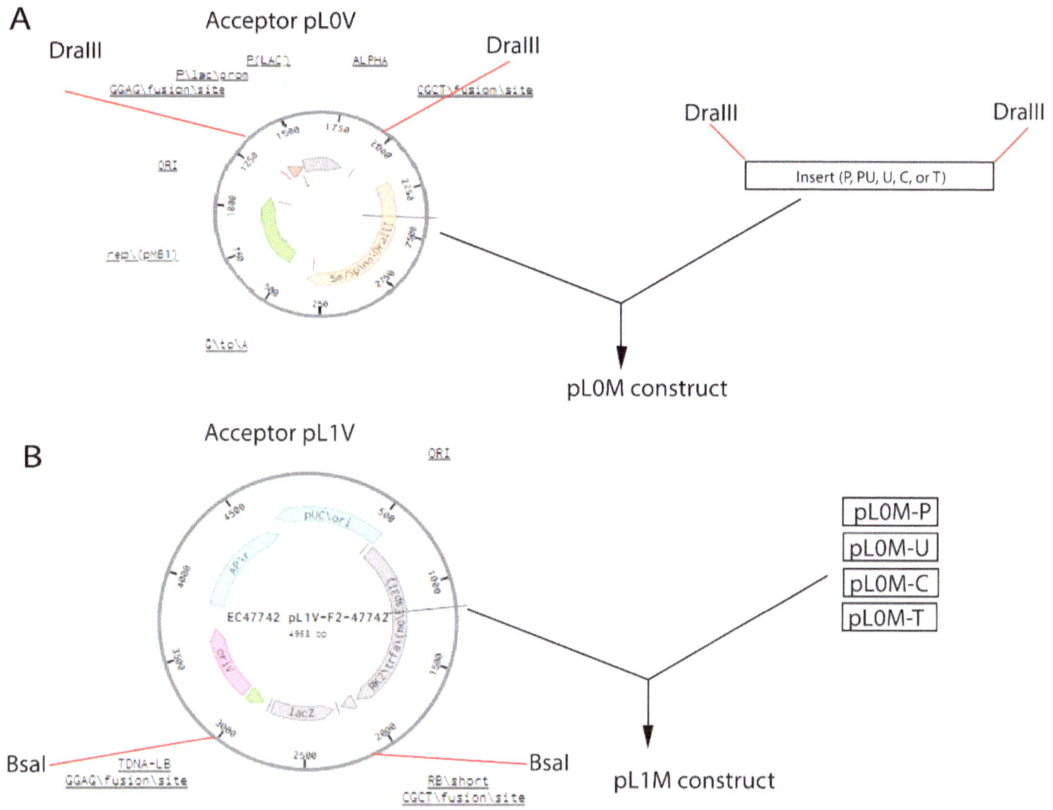

Fig. 3 An overview of the modular Golden Gate cloning and assembly. (**a**) Synthesized genetic elements like promoter (P), 5; untranslated region or UTR (U), coding region (C), terminator (T) and cloned into pL0V (acceptor plasmid or vector) using DraII restriction enzyme. Different classes of pL0M constructs are flanked by compatible unique fusion sites consisting of 4 nucleotides and a Type IIS enzyme recognition site (BsaI). (**b**) transcriptional units containing multiple genes selected from the pL0M constructs are assembled using one-pot-one step cloning reaction to generate level 1 constructs (pL1M)

3. To obtain L1 plasmids, combine L0 modules with transcriptional expression units that contain the promotor—UTR (PU) coding sequence (SC) and terminator (T) modules. Use Type II restriction enzymes such as BasI-HF and T4 Ligase for assembly of L1 plasmids (Fig. 3b).
4. To combine more than one gene into a single plasmid, create L2 plasmids using L1 modules and BbsI-HF and T4 Ligase enzymes.
5. To ensure correct assembly, all constructs (L0, and L1) are confirmed by using a diagnostic digestion followed by sequencing the end-junctions of the inserts.

3.4 Agrobacterium tumefaciens *Transformation*

1. Transform the assembled L1 and L2 plasmids into *Agrobacterium tumefaciens* strain GV3101 using transformation by electroporation [20].
2. Incubate 500 ng of L1 or L2 plasmid with electrocompetent *Agrobacterium tumefaciens* strain GV3101.
3. Upon electroporation, rescue the transformed *Agrobacterium tumefaciens* by adding 1 mL of LB shaking at 28 °C for 1.5–2 h.
4. Plate transformants onto LB agar that contains the required antibiotics.
5. Perform colony PCR on a few selected colonies of Agrobacterium before setting up liquid overnight culture of a single selected colony.
6. Use one gene-specific primer in conjunction to vector specific primer for probing presence of inserted gene.
7. Select and inoculate single colonies after incubating for 2 days at 28 °C for transiently expressing the gene of interest in tobacco leaf.

3.5 *Transient Expression in Tobacco Leaf Epidermal Cell*

1. Grow transformed Agrobacterium containing the desired constructs by growing the cells at 28 °C overnight in 5 mL of LB that contains the appropriate antibiotics and shaking at 220 rpm.
2. Pellet the overnight grown bacterial culture by centrifuging at 10,000 rpm for 1 min at room temperature.
3. Resuspended bacterial pellet in 1 mL of Agrobacterium Infiltration Medium and incubate for 2 h with shaking at 150 rpm.
4. Maintain the target strains at an optical density of OD600 = 0.5 per strain and internal control plasmids, namely, Nano-LUC and p19 at OD600 = 0.3.
5. Co-infiltrate the plasmid harboring gene of interest and control plasmids (*see* **Notes 8**, **9**, and **10**).
6. Maintain the infiltrated leaves under ambient conditions post-infiltration. After 3 days, harvest the infiltrated leaves for further analysis.
7. In the case of variable conditions, transfer the infiltrated tobacco plants to the appropriate stress conditions 2 days after infiltration (as outlined below) and then harvest the leaves for further analysis after 1 day.

3.6 In Vitro Luciferase Assay

For determining the efficacy of CREs, the Dual Luciferase Reporter Assay System is used following manufacturer's instructions.

1. Collect leaf discs of 6 mm diameter from infiltrated leaf are homogenized in 150 μL of extraction buffer (*see* **Note 11**).
2. Pipette 20 μL of crude extract and mix with 75 μL of LUC assay buffer and add to an opaque 96-well plate.
3. Monitor LUC activity spectrophotometrically using a multimode plate reader.
4. Next, add 75 μL of stop and glow buffer for the measurement of the *Renilla* luciferase activity.
5. Monitor sequentially the firefly luciferase and *Renilla* luciferase to calculate the Fluc:Rluc ratio.
6. Normalize the data as the ratio of the luminescent signal intensity for reporter versus internal control reporter (pNOS) from three independent biological samples.

3.7 Abiotic Stress Treatment

3.7.1 Low CO_2 Stress Treatment

To determine the effect of low CO_2 on the efficacies of synthetic promoters containing photorespiratory CREs, follow the steps below:

1. Post 2 days of agroinfiltration, transfer one set of agroinfiltrated plants to low CO_2 (200 ppm) and a temperature regime of 22 °C.
2. Maintain the second set of plants (as control) at 400 ppm CO_2 and a temperature regime of 22 °C. For each set, 6 plants were used.
3. Subject the agroinfiltrated plants to low CO_2 for 24 h.
4. Harvest the leaf tissue by flash freezing in liquid nitrogen and perform luciferase assay outlined above (Fig. 4).

3.7.2 High-Temperature Stress Treatment

To determine the effect of high temperature on the efficacies of synthetic promoters containing photorespiratory CREs, follow the steps below:

1. Post 2 days of agroinfiltration, transfer one set of agroinfiltrated plants to elevated temperature of 32 °C. For each set, 6 plants were used.
2. Maintain the second set of plants (as control) at a temperature regime of 22 °C.
3. Subject the agroinfiltrated plants to high temperature for 1 h and 3 h.
4. Harvest the leaf tissue by flash freezing in liquid nitrogen and perform luciferase assay as outlined above (Fig. 5).

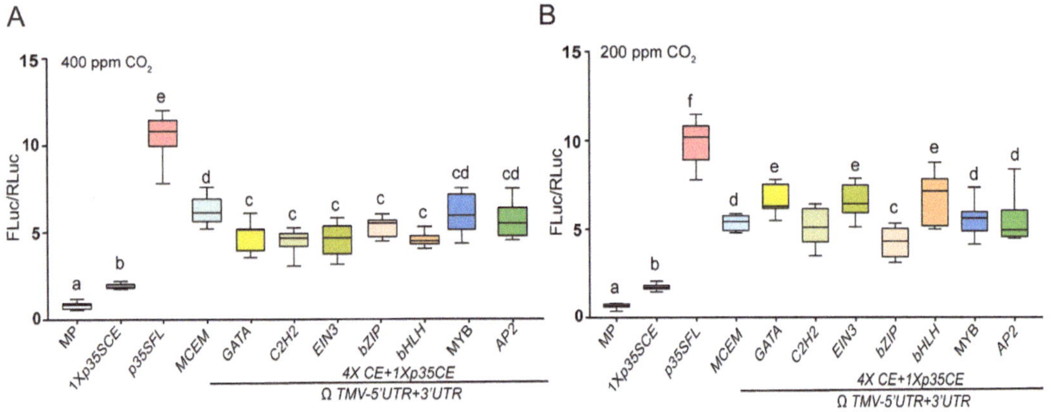

Fig. 4 Effect of low CO_2 on the efficacies of synthetic promoters derived from CREs of core photorespiratory genes. (**a**) Relative luciferase activity driven by synthetic promoters under ambient condition (400 ppm CO_2). (**b**) Relative luciferase activity driven by synthetic promoters under low CO_2 (200 ppm CO_2). Constructs were agroinfiltrated and 2-day post infiltration one set of transformed plants were shifted to a growth chamber with low CO_2 (200 ppm) for 24 h and control plants were kept at 400 ppm of CO_2. The UTRs used are indicated by the line at the bottom. Luciferase activity within the infiltrated tobacco epidermal cells was quantified by measuring light emission in a luminometer. Promoter activity is expressed as the relative activity of firefly luciferase versus *Renilla* luciferase (Fluc/Rluc). Box-whiskers plots represent data from three independent experiments, each with three biological replicates. Error bar represents a mean ± standard deviation. Letters indicate statistical significance determined by one-way ANOVA with Tukey's multiple component analysis ($p < 0.05$); samples sharing letters are not significantly different

3.7.3 High Light Intensity Stress Treatment

To determine the effect of high light intensity on the efficacies of synthetic promoters containing photorespiratory CREs, follow the steps below:

1. Post 2 days of agroinfiltration, transfer one set of agro-infiltrated plants to high light intensity of 1200 μmol photons m^{-2} s^{-1} and temperature regime of 25 °C.
2. Maintain the second set of plants (as a control) at 150 μmol photons m^{-2} s^{-1} temperature regime of 25 °C. For each set 6 plants were used.
3. Subject the agroinfiltrated plants to high light intensity for 3 h.
4. Harvest the leaf tissue by flash freezing in liquid nitrogen and perform luciferase assay as outlined above (Fig. 6).

4 Notes

1. Most often "inducible" in molecular biology usually refers to a chemical or optical induction of a gene expression, it can also be something induced by a signal in nature as well. For instance, stress-responsive promoters turn on or off in response to particular stress.

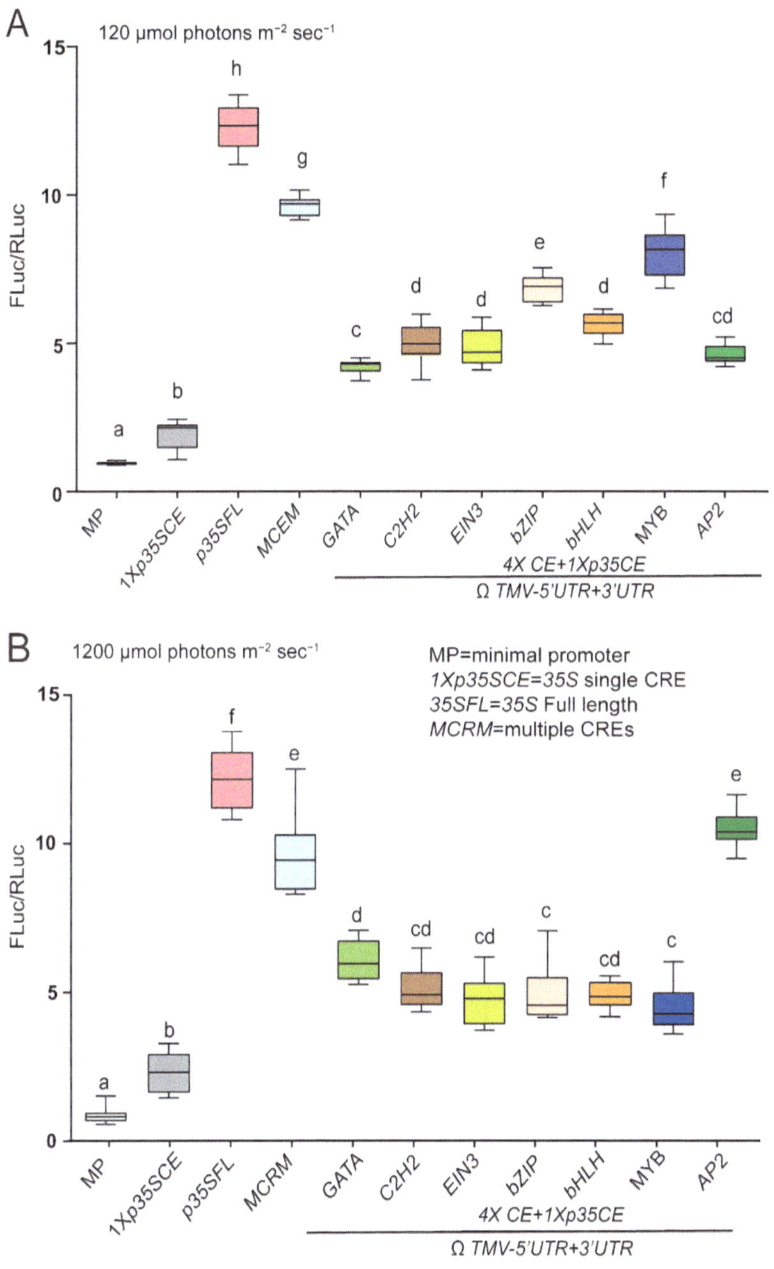

Fig. 5 Effect of high light intensity on the efficacies of synthetic promoters derived from CREs of core photorespiratory genes. (**a**) Relative luciferase activity driven by synthetic promoters under ambient condition (150 μmol photons m^{-2} s^{-1}). (**b**) Relative luciferase activity driven by synthetic promoters under ambient condition (1200 μmol photons m^{-2} s^{-1}). Constructs were agroinfiltrated and 2-day post infiltration one set of transformed plants were subjected to light intensity 1200 μmol photons m^{-2} s^{-1} and control transformed plants were maintained at 150 μmol photons m^{-2} s^{-1}. The UTRs used are indicated by the line at the bottom. Data analyzed as described in Fig. 3. Box-whiskers plots represent data from three independent experiments, each with three biological replicates. Error bar represents a mean ± standard deviation. Letters in the indicate statistical significance based on a one-way ANOVA with Tukey's multiple component analysis ($p < 0.05$); samples sharing letters are not significantly different

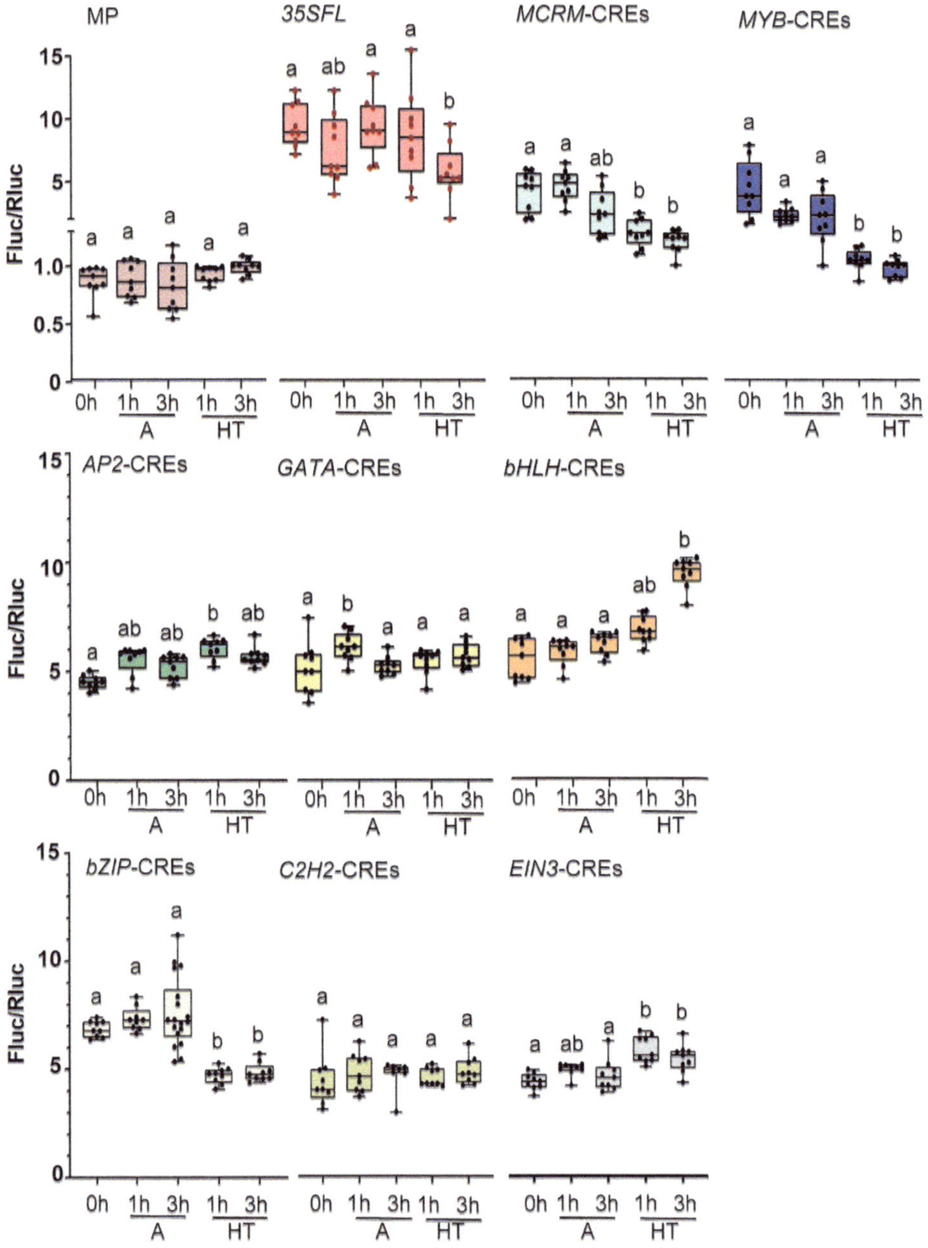

Fig. 6 Time-resolved luciferase activity displaying the impact of elevated temperature on efficacies of native CREs. 'A' denotes ambient conditions (22 °C) and HT denotes high temperature (32 °C) for indicated time-points. Constructs were agroinfiltrated and 2-day post infiltration one set of transformed plants were shifted to a growth chamber with elevated temperature (32 °C) for 1 h, and 3 h. Data analyzed as described in Fig. 3. Box-whiskers plots represent data from three independent experiments, each with three biological replicates. Error bar represents a mean ± standard deviation. Letters in the indicate statistical significance based on a one-way ANOVA with Tukey's multiple component analysis ($p < 0.05$); samples sharing letters are not significantly different

2. The ambient and low level of CO_2 is monitored using a sensor and measured amount of soda lime (to reduce CO_2) and CO_2.
3. For selecting common stress-responsive cis-regulatory elements shared by PR genes, set stringent parameters in the two web-based tools. For instance, while using PlantPAN3.0, the threshold for co-occurrence support was kept at >90% confidence and distance constraint at 100 bp.
4. The L0 constructs were synthesized in collaboration with Twist Biosciences or Engineering Nitrogen Symbiosis for Africa ENSA group (University of Cambridge, Cambridge, United Kingdom) [17]. It is important to select either Kan^R or $Spec^R$ antibiotics as selectable markers.
5. For Golden Gate assembly, 2:1 molar ratio of insert:vector was used.
6. The reaction mixture for assembling L0 and L1 modules was incubated for 2 min at 37 °C, 5 min at 16 °C. Both these two steps were repeated 45× times, followed by incubation for 5 min at 50 °C and 10 min at 80 °C in a thermocycler.
7. T4 DNA ligase buffer needs to be aliquoted as it contains ATP, which is heat labile. Thus, repeated freeze-thaw cycles should be avoided.
8. Tobacco plants grown for 4- to 6-week-old were used for transient assay as those leaves are soft enough for uniform infiltration.
9. Infiltration of tobacco leaves was performed at the abaxial surface of tobacco leaves using 1ml needle-less syringe.
10. Avoid the mid-vein of the leaf while infiltrating as higher resistance may be experienced.
11. For each luciferase assay, 2–3 leaf discs per infiltrated leaves were used.

Acknowledgments

This work was supported by the research project Realizing Increased Photosynthetic Efficiency (RIPE) that is funded by the Bill & Melinda Gates Foundation, Foundation for Food and Agriculture Research (FFAR), and the UK Foreign Commonwealth & Development Office under grant number OPP1172157.

References

1. Anderson LE (1971) Chloroplast and cytoplasmic enzymes. II. Pea leaf triose phosphate isomerases. Biochimica et biophysica acta 235: 237–244
2. Peterhansel C et al (2010) Photorespiration. The arabidopsis book 8:e0130
3. Peterhansel C et al (2013) Engineering photorespiration: current state and future

possibilities. Plant Biol (Stuttgart, Germany) 15:754–758

4. South PF, Cavanagh AP, Lopez-Calcagno PE, Raines CA, Ort DR (2018) Optimizing photorespiration for improved crop productivity. J Integr Plant Biol 60:1217–1230
5. Shi X, Bloom A (2021) Photorespiration: the futile cycle? vol 10, Plants (Basel, Switzerland)
6. Bloom AJ (2015) Photorespiration and nitrate assimilation: a major intersection between plant carbon and nitrogen. Photosynth Res 123:117–128
7. Abadie C, Tcherkez G (2019) Plant sulphur metabolism is stimulated by photorespiration. Commun Biol 2:379
8. Keech O, Gardeström P, Kleczkowski LA, Rouhier N (2017) The redox control of photorespiration: from biochemical and physiological aspects to biotechnological considerations. Plant Cell Environ 40:553–569
9. Voss I, Sunil B, Scheibe R, Raghavendra AS (2013) Emerging concept for the role of photorespiration as an important part of abiotic stress response. Plant Biol 15:713–722
10. Sharkey TD (1988) Estimating the rate of photorespiration in leaves. Physiologia Plantarum 73:147–152
11. Slattery RA, Ort DR (2019) Carbon assimilation in crops at high temperatures. Plant Cell Environ 42:2750–2758
12. Huang W, Hu H, Zhang S-B (2015) Photorespiration plays an important role in the regulation of photosynthetic electron flow under fluctuating light in tobacco plants grown under full sunlight. Front Plant Sci 6:621
13. Fu X, Walker BJ (2023) Dynamic response of photorespiration in fluctuating light environments. J Exp Bot 74:600–611
14. Somerville CR, Ogren WL (1979) A phosphoglycolate phosphatase-deficient mutant of Arabidopsis. Nature 280:833–836
15. Pick TR et al (2013) PLGG1, a plastidic glycolate glycerate transporter, is required for photorespiration and defines a unique class of metabolite transporters. Proc Natl Acad Sci U S A 110:3185–3190
16. South PF et al (2017) Bile Acid Sodium Symporter BASS6 can transport glycolate and is involved in photorespiratory metabolism in Arabidopsis thaliana. Plant Cell 29:808–823
17. Basu D, South PF (2022) Design and analysis of native photorespiration gene motifs of promoter untranslated region combinations under short term abiotic stress conditions. Front Plant Sci 13
18. Weber E, Gruetzner R, Werner S, Engler C, Marillonnet S (2011) Assembly of designer TAL effectors by golden gate cloning. PLoS One 6:e19722
19. Engler C et al (2014) A Golden Gate modular cloning toolbox for plants. ACS Synth Biol 3: 839–843
20. Nagel R, Elliott A, Masel A, Birch RG, Manners JM (1990) Electroporation of binary Ti plasmid vector into Agrobacterium tumefaciens and Agrobacterium rhizogenes. FEMS Microbiol Lett 67:325–328

Chapter 21

Identification of Auxiliary Organellar Targeting Signals for Plant Peroxisomes Using Bioinformatic Analysis of Large Protein Sequence Datasets Followed by Experimental Validation

Qianwen Deng, Hangjin Jiang, Jianping Hu, and Ronghui Pan

Abstract

Eukaryotic cells are compartmentalized by membrane-bounded organelles to ensure that specific biochemical reactions and cellular functions occur in a spatially restricted manner. The subcellular localization of proteins is largely determined by their intrinsic targeting signals, which are mainly constituted by short peptides. A complete organelle targeting signal may contain a core signal (CoreS) as well as auxiliary signals (AuxiS). However, the AuxiS is often not as well characterized as the CoreS. Peroxisomes house many key steps in photorespiration, besides other crucial functions in plants. Peroxisome targeting signal type 1 (PTS1), which is carried by most peroxisome matrix proteins, was initially recognized as a C-terminal tripeptide with a "canonical" consensus of [S/A]-[K/R]-[L/M]. Many studies have shown the existence of auxiliary targeting signals upstream of PTS1, but systematic characterizations are lacking. Here, we designed an analytical strategy to characterize the auxiliary targeting signals for plant peroxisomes using large datasets and statistics followed by experimental validations. This method may also be applied to deciphering the auxiliary targeting signals for other organelles, whose organellar targeting depends on a core peptide with assistance from a nearby auxiliary signal.

Key words Auxiliary targeting signal, Large dataset, Organelle, Peroxisome Targeting Signal type 1 (PTS1), Peroxisome, Protein subcellular localization, Statistical analysis

1 Introduction

Eukaryotic cells are compartmentalized by membrane-bounded organelles to ensure that specific biochemical reactions and cellular functions occur in a spatially restricted manner [1]. Each of the major organelles present in eukaryotic cells, including the nucleus, endoplasmic reticulum, mitochondria, plastids, golgi apparatus, lysosomes, endosomes, and peroxisomes, contains a distinct suite

Berkley J. Walker (ed.), *Photorespiration: Methods and Protocols*, Methods in Molecular Biology, vol. 2792,
https://doi.org/10.1007/978-1-0716-3802-6_21,

of proteins involved in specialized biological functions. The sorting of intracellular proteins into the organelles is generally dictated by short peptide sequences at or near the C- or N-terminus of a nascent protein, which differ in their length, structure, and position in the protein [2–5].

The single membrane-delimited peroxisome is an essential organelle to many eukaryotes, as manifested by the serious human genetic diseases and defects in plant growth and development displayed in peroxisomal mutants [6–8]. Plant peroxisomes house many crucial functions, such as photorespiration, fatty acid degradation, the glyoxylate cycle, phytohormone biosynthesis, and catabolism of reactive oxygen species (ROS) [9–11]. The functional diversity, complexity, and species-specificity of plant peroxisomes are accomplished by the large amount and variety of proteins in these organelles [10, 12–14]. The sorting of its matrix proteins relies on peroxisome targeting signal type 1 (PTS1) at the C-terminus and PTS2 near the N-terminus of the protein [10, 15].

PTS1 is found in most peroxisome matrix proteins and was initially recognized as a C-terminal tripeptide (PTS1-3aa) with a "canonical" consensus of [S/A]-[K/R]-[L/M]. Later, more and more "non-canonical" PTS1s were discovered, demonstrating the complexity of PTS1 [16–19]. It is apparent that proteins containing the canonical PTS1-3aa as its core signal (CoreS) are more likely to direct the protein to peroxisomes than those containing the noncanonical ones. However, in many cases, even the major canonical PTS1-3aa SKL> is not sufficient to determine peroxisome localization by itself [20]. A number of studies have shown that 9 to 11 amino acids (aa) upstream of the PTS1-3aa can significantly affect the peroxisome targeting ability of PTS1-3aa [15, 17, 21–25], strongly suggesting the existence of auxiliary signals (AuxiS) in PTS1-mediated peroxisome targeting.

To fully understand the function of organelles, it is essential to better understand how proteins are sorted to these organelles. In a recent study, we used large datasets retrieved from 177 plant genomes to perform a statistics-based systematic analysis of the PTS1 domain (CoreS plus its upstream sequence), which resulted in the successful identification of AuxiS for plant PTS1 [20]. Here, we describe the method employed, which starts with the generation of a large dataset of proteins across diverse plant species, using established peroxisomal and non-peroxisomal proteins for BLAST-based homolog searches, followed by bioinformatic analysis (Fig. 1) and complemented by experimental validation. We expect that this method can be applied to identifying AuxiS for other organelles whose protein targeting is also dependent on a CoreS with the assistance of an AuxiS nearby.

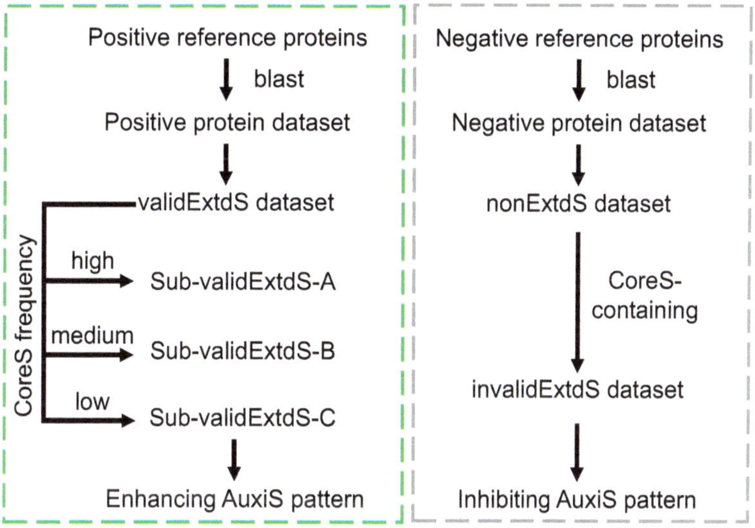

Fig. 1 Workflow for assembling the peptide datasets and predicting enhancing and inhibiting AuxiS patterns. CoreS indicates core signal. AuxiS indicates auxiliary signal. Green and gray dotted boxes represent the dataset generation processes for analyzing the enhancing and inhibiting AuxiS patterns, respectively

2 Materials

2.1 Vector Construction

1. pCAMBIA1300-mVenus vector.
2. *Xba* I and *Sac* I restriction endonucleases.
3. High-fidelity DNA polymerase.
4. PCR instrument.
5. Electrophoresis buffer: 50× TAE buffer is prepared by mixing and dissolving 242 g Tris base, 100 mL 0.5 M EDTA, and 57.1 mL glacial acetic acid in deionized water to a final volume of 1 L. The pH of the final solution is between 8.2 and 8.4. 1× TAE buffer is used for electrophoresis.
6. 1% agarose gel: 1g agarose per 100 mL final volume of electrophoresis buffer.
7. Lysogeny broth (LB) medium (liquid and solid): 1% bacto-tryptone, 0.5% yeast extract, and 1% NaCl for liquid LB medium. An extra 1.5% agar powder is added to make solid LB medium.
8. DH5α competent *Escherichia coli* cells.
9. GV3101(pMP90) chemically competent *Agrobacterium tumafaciens* cells.

2.2 Transient Protein Expression in Tobacco Leaves

1. Peroxisome marker moxCerulean3-PTS1: coding region of the moxCerulean3 fluorescent protein fused with a SKL tripeptide to its C-terminus is cloned into the pGWB545 vector backbone. Agrobacterium cells containing the construct with spectinomycin (spe) and rifampicin (rif) resistance are obtained by transforming the construct into Agrobacterium GV3101 (pMP90).
2. Centrifuge.
3. 1 mL injector.
4. Tobacco: Four-week-old *Nicotiana tabacum*.

2.3 In Vivo Protein Targeting Analysis

1. Slides and coverslip.
2. Distilled water.
3. Confocal laser-scanning microscope.

3 Methods

3.1 Collection of Protein Sequences

1. Check for angiosperm plant species with assembled genomes on the website of Published Plant Genomes (https://www.plabipd.de/).
2. Download the proteome sequences of angiosperm plant species from the following genome databases:
   (a) NCBI Genome (https://www.ncbi.nlm.nih.gov/genome).
   (b) Phytozome v13 (https://phytozome.jgi.doe.gov) [26].
   (c) Ensembl Plants release 49 (https://plants.ensembl.org) [27].
3. Search the literature for established peroxisomal proteins with a CoreS, i.e., Arabidopsis PTS1-3aa-containing proteins previously validated to be peroxisomal or known to be involved in a well-established peroxisomal pathway (*see* **Note 1**). Use these CoreS-containing proteins as positive reference proteins (Fig. 1).
4. Prepare a list of established transcription factor (TF) proteins (e.g., Arabidopsis proteins from the MYB, bZIP, ARF, NAC, WRKY, and MADS families) as non-peroxisomal negative reference proteins (Fig. 1).
5. Perform sequence homology-based searches in the angiosperm plant species identified in **step 1** for mutual best-match proteins of the positive (peroxisomal) and negative (non-peroxisomal) reference proteins (*see* **Note 2**). Use the mutual best-match protein lists as positive or negative protein datasets for follow-up analysis (Fig. 1).

3.2 Assembly of Peptide Datasets

1. Improve the quality of the positive protein dataset by removing proteins with less conserved CoreS, i.e., proteins containing C-terminal tripeptide (C-3aa) that appeared in <10 proteins in the positive protein dataset (*see* **Note 3**).
2. Extract the extend peptides (ExtdS) containing both the CoreS and AuxiS, e.g., the C-terminal 15 aa (C-15aa), from the positive protein dataset. The nonredundant ExtdS are kept for the validExtdS (e.g., validPTS1-15aa) dataset (Fig. 1).
3. Check for novel CoreS in the validExtdS dataset and experimentally validate their targeting function. For each novel CoreS, choose 2 to 3 ExtdS peptides for experimental validation (*see* **Note 4**) using transient expression in tobacco leaves (*see* Subheading 3.3).
4. Extract ExtdS from the negative protein dataset. Keep the nonredundant ExtdS for the nonExtdS (e.g., nonPTS1-15aa) dataset (Fig. 1).
5. Select all ExtdS that contain a validated CoreS (e.g., PTS1-3aa) from the nonExtdS dataset to form the invalidExtdS (e.g., invalidPTS1-15aa) dataset (*see* **Note 5**) (Fig. 1).

3.3 Experimental Validation of the ExtdS

1. Convert the selected ExtdS into nucleotide sequences (e.g., 45 bp for PTS1-15aa) based on codon preference of the corresponding reference species provided on the website http://www.molbiol.ru/eng/scripts/01_19.html.
2. Design three primers (see below) using the scheme depicted in Fig. 2:
   (a) Primer F contains 15 bp sequence before the *Xba* I cleavage site of the pCAMBIA1300-mVenus vector and 15 bp encoding the first 5 aa of mVenus.
   (b) Primer R1 contains 36 bp sequence encoding the first 12 aa of the target peptides and 15 bp encoding the last 5 aa of mVenus.
   (c) Primer R2 contains 15 bp sequence after the *Sac* I cleavage site of the pCAMBIA1300-mVenus vector, the 3-bp stop codon, and 30 bp encoding the last 10 aa of the target peptides.
3. Using the pCAMBIA1300-mVenus vector as template and F and R1 as primers, amplify the fragment containing a fusion of mVenus and the first 12 aa of the target peptides.
4. Using the PCR product from the previous step and F and R2 as primers, amplify a fragment containing the complete target fragment with the 15 bp sequence before the *Xba* I cleavage site and 15 bp after the *Sac* I cleavage site of the pCAMBIA1300-mVenus vector at both ends. Extract the PCR fragment after 1% agarose gel electrophoresis.

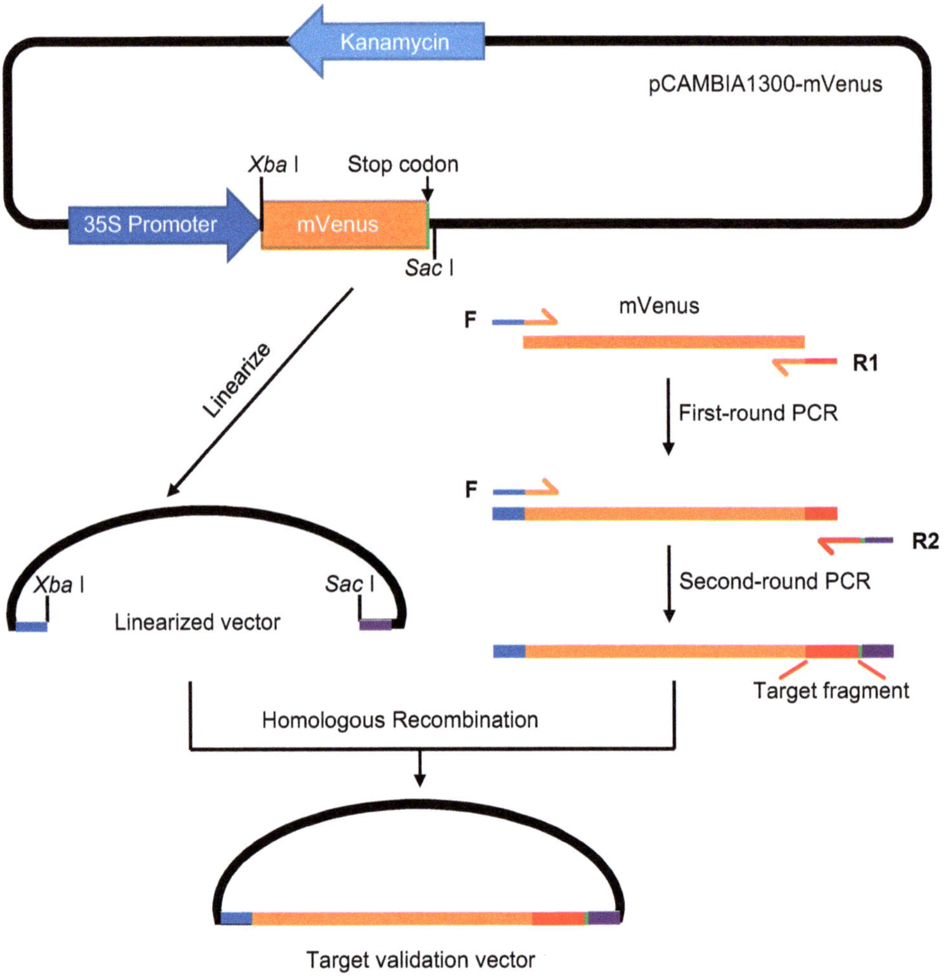

Fig. 2 Procedure for constructing the target validation vector. Black lines represent the vector backbone of pCAMBIA1300-mVenus. Orange lines indicate coding sequence for the mVenus fluorescent protein. Blue lines are the 15 bp sequence before the *Xba* I cleavage site of the pCAMBIA1300-mVenus vector. Purple lines represent the 15 bp sequence after the *Sac* I cleavage site of the pCAMBIA1300-mVenus vector. Red lines represent sequence encoding target peptides. Green lines indicate the stop codon

5. Digest the pCAMBIA1300-mVenus vector with *Xba* I and *Sac* I, before connecting it with the fusion fragment through homologous recombination. Transfer the final product into DH5α cells by heat shock.
6. After sequencing, transform the correct construct into Agrobacterium GV3101 (pMP90) by freeze-thaw [28] and select for target Agrobacterium cells with kanamycin (kan) and rif resistance.
7. Incubate the target-containing Agrobacterium or the peroxisome marker-containing Agrobacterium in LB supplemented with the corresponding antibiotics at 28 °C with 200 rpm

shaking for 14~16 h. Then, centrifuge the culture for 5 min at 5000 rpm, discard the supernatant, and resuspend the sediment with sterile water in the same volume as the discarded supernatant.

8. Mix the resuspended target and marker Agrobacterium at 1:1 in volume. Inject the mixture into leaves of 4-week-old tobacco plants with 1 ml injector. The target Agrobacterium is separately injected as a control.
9. 72 h after infiltration, use confocal laser-scanning microscope to capture images. mVenus is excited with 514-nm lasers and detected at 530–630 nm and moxCerulean3 is excited with 445-nm lasers and detected at 460–500 nm. Overlapping of the signals from the target fusion protein and the organelle marker indicates that the selected ExtdS is functional in peroxisome targeting.

3.4 Assessment of the Relation Between the AuxiS and CoreS

1. Categorize the validExtdS dataset into three subsets based on the occurrence frequency of CoreS (*see* **Notes 6** and 7).
   (a) Sub-validExtdS-A: CoreS with high frequency of occurrence (e.g. $\geq$ 200 in the validPTS1-15aa dataset).
   (b) Sub-validExtdS-B: CoreS with medium frequency of occurrence (e.g. between 100 and 200 in the validPTS1-15aa dataset).
   (c) Sub-validExtdS-C: Low-frequency CoreS (e.g., <100 in the validPTS1-15aa dataset).
2. Calculate the Kullback–Leibler (KL) divergence (*see* **Note** 7) for the extended region of ExtdS (e.g., each position of the 12-aa sequence upstream of PTS1-3aa), using peptides from the three sub-validExtdS datasets against those from the non-ExtdS dataset. Then determine the correlation between the frequency of the CoreS and their upstream KL values. Sub--validExtdS-C, which contains CoreS with the lowest frequency of occurrence, should show the maximum KL divergence. This is because lower frequency equals to lower evolutionary selection power and weaker targeting ability. A weaker CoreS is more dependent on a stronger upstream auxiliary signal.
3. Generate seqlogo plots (*see* **Note 8**) for the extended region of ExtdS using the ExtdS of the sub-validExtdS-C dataset against those of the nonExtdS dataset. Deduce an enhancing AuxiS pattern using the most enriched residue at each position.
4. Calculate KL divergence and generate seqlogo plot of the extended region of ExtdS using peptides from the invalidExtdS dataset against those from the validExtdS dataset. Then deduce an upstream inhibiting AuxiS pattern using the most enriched residue at each position.

5. Fuse these deduced upstream enhancing or inhibiting AuxiS patterns with several selected CoreS. Check the ability of these AuxiS patterns to impact the organelle targeting efficiency of CoreS using transient protein expression in tobacco (*see* Subheading 3.3). Using different time points, such as 36 h, 48 h, and 72 h post infiltration, classify the targeting efficiency as No, Complete, or Incomplete.

3.5 Effects of Residue Polarity and Charge in the AuxiS

1. Group the amino acid residues according to their polarity and charge, and regenerate KL divergence and seqlogo plots of the extended region of ExtdS for the sub-validExtdS datasets and the invalidExtdS dataset to reveal the importance of residue polarity and charge.
   (a) Basic: R, K, and H, as letter B in the seqlogo plots
   (b) Acidic: E and D, as letter X
   (c) Nonpolar: I, P, F, W, A, V, L, and M, as letter U
   (d) Polar neutral: C, Y, N, G, S, T, and Q, as letter O
2. Based on the seqlogo plot of the sub-validExtdS-C dataset in **step 3** of Subheading 3.4, determine which types of residues are enriched in the sub-validExtdS-C dataset. Deduce peptides for the extended region of ExtdS using the most enriched residues of the enriched types at each position (e.g., an all-basic 12 aa peptide and an all-nonpolar 12 aa peptide for validPTS1-15aa revealed by Deng et al. [20]).
3. Based on the seqlogo plot of the invalidExtdS dataset in **step 4** of Subheading 3.4, determine which types of residues are enriched in the invalidExtdS dataset. Deduce peptides for the extended region of ExtdS using the most enriched residues of the enriched types at each position (e.g., an all-acidic 12 aa peptide and an all-polar neutral 12 aa peptide for invalidPTS1-15aa revealed by Deng et al. [20]).
4. Determine the effects of residue polarity and charge in the AuxiS on the organelle targeting ability of selected CoreS, using tobacco transient expression analysis (*see* Subheading 3.3) at different time points. Use the deduced peptides in **steps 2** and **3** to represent different patterns of polarity and charge at the extended region of ExtdS.

3.6 The Core Position and Pattern of the AuxiS

1. Compare the seqlogo plot of the extended region of ExtdS between the sub-validExtdS-C dataset against nonExtdS dataset and invalidExtdS dataset against validExtdS dataset. Determine the core residue positions, which show the biggest difference.
2. Perform residue exchanges between the enhancing and inhibiting AuxiS patterns for the residues at the core positions, and generate hybrid extended regions of ExtdS peptides.

3. Check the impacts of these hybrid extended regions of ExtdS on the organelle targeting ability of selected CoreS, using tobacco transient expression analysis (*see* Subheading 3.3) at different time points. Residues at the core positions should have the deciding impact on peroxisome targeting.

4 Notes

1. Since Arabidopsis proteins that contain both PTS1 and PTS2 are less conserved across species, proteins containing both types of PTS are removed from reference proteins [20].
2. Search for mutual homologs using DIAMOND v2.0.6 with the sensitive model and e value 1×10^{-10} [29] to identify the mutual best match for each reference protein. A protein is selected only if it is the "mutual best-match" in the two-way BLAST search between its corresponding plant species and reference protein source species Arabidopsis. Protein sequence identity of 30% is the threshold for a protein to be considered.
3. The smaller the value, the more sequences are needed to be verified in vivo and the larger the dataset. Therefore, a moderate filter value of 10 or 5 is usually chosen.
4. It is best to select ExtdS peptides derived from multiple reference proteins for validation. If the validation results of different ExtdS peptides containing the same CoreS are inconsistent, all ExtdS peptides containing this CoreS should be tested.
5. The invalidExtdS dataset should include as many canonical CoreS as possible, otherwise the deduced upstream inhibiting AuxiS pattern may not be strong enough.
6. The optimal partition value refers to the different degrees of KL divergence between three sub-datasets. The optimal partition value is the partition value when the KL divergence difference between the three sub-datasets is the largest.
7. KL divergence, also known as Relative Entropy, quantifies how much one probability distribution differs from another probability distribution; for example, a KL value of 0 indicates no difference [30]. We used two discrete probability distributions, $P(z)$ and $Q(z)$, to represent the distribution of amino acids at each position of the sequence. KL divergence from Q to P is defined as:

$$D_{\mathrm{KL}}\left(P(z)\middle\|Q(z)\right) \equiv \sum_{z} P(z) \log\left[\frac{P(z)}{Q(z)}\right]$$

Because KL divergence is not symmetrical, which means that $D_{KL}(P \| Q) \neq D_{KL}(Q \| P)$, we used a symmetric KL divergence to compare the distribution of amino acids at each position defined as $\frac{1}{2} D_{KL}\left(P \| Q\right) + \frac{1}{2} D_{KL}\left(Q \| P\right)$ [20].

8. A seqlogo consists of a stack of amino acids represented by single letters at each position. The height of the letters represents the information content at this position, whereas bigger-sized letters indicate lower variability, higher evolutionary conservation, and higher information content [31]. The information content of position i is calculated as:

$$I_i = \log_2 20 - H_i$$

H_i is the Shannon entropy of position i. 20 indicates the total types of amino acids. $f_{b,\ i}$ is defined as the relative frequency of amino acid b at position i. H_i is calculated as follows:

$$H_i = -\sum_{b=1}^{20} f_{b,i} \times \log_2 f_{b,i}$$

Therefore, the height of amino acid b in column i is calculated as $f_{b,\ i} \times I_i$ [20].

Acknowledgment

This work was supported by the Zhejiang Provincial Natural Science Foundation of China (LZ23C020002), the National Natural Science Foundation of China (32200231), the National Key Research and Development Program (2022YFD1401600), and Key Research Project of Zhejiang Lab (2021PE0AC04) to R.P., and by the National Science Foundation (MCB 2148206) to J.H.

References

1. Zhao YG, Zhang H (2020) Phase separation in membrane biology: the interplay between membrane-bound organelles and membraneless condensates. Dev Cell 55:30–44
2. Teufel F, Almagro Armenteros JJ, Johansen AR et al (2022) SignalP 6.0 predicts all five types of signal peptides using protein language models. Nat Biotechnol 40:1023–1025
3. Chu CC, Swamy K, Li HM (2020) Tissue-specific regulation of plastid protein import via transit-peptide motifs. Plant Cell 32: 1204–1217
4. Murcha MW, Kmiec B, Kubiszewski-Jakubiak S et al (2014) Protein import into plant mitochondria: Signals, machinery, processing, and regulation. J Exp Bot 65:6301–6335
5. Emanuelsson O, Brunak S, von Heijne G et al (2007) Locating proteins in the cell using TargetP, SignalP and related tools. Nat Protoc 2: 953–971
6. Hu J, Baker A, Bartel B et al (2012) Plant peroxisomes: biogenesis and function. Plant Cell 24:2279–2303
7. Pan R, Liu J, Hu J (2019) Peroxisomes in plant reproduction and seed-related development. J Integr Plant Biol 61:784–802
8. Honsho M, Okumoto K, Tamura S et al (2020) Peroxisome biogenesis disorders. Adv Exp Med Biol 1299:45–54
9. Li J, Tietz S, Cruz JA et al (2019) Photometric screens identified Arabidopsis peroxisome

proteins that impact photosynthesis under dynamic light conditions. Plant J 97:460–474

10. Pan R, Liu J, Wang S et al (2020) Peroxisomes: versatile organelles with diverse roles in plants. New Phytol 225:1410–1427
11. Midorikawa K, Tateishi A, Toyooka K et al (2022) Three-dimensional nanoscale analysis of light-dependent organelle changes in Arabidopsis mesophyll cells. PNAS Nexus 1:1–9
12. Reumann S, Bartel B (2016) Plant peroxisomes: recent discoveries in functional complexity, organelle homeostasis, and morphological dynamics. Curr Opin Plant Biol 34:17–26
13. Charton L, Plett A, Linka N (2019) Plant peroxisomal solute transporter proteins. J Integr Plant Biol 61:817–835
14. Corpas FJ (2019) Peroxisomes in higher plants: an example of metabolic adaptability. Bot Lett 166:298–308
15. Reumann S (2004) Specification of the peroxisome targeting signals type 1 and type 2 of plant peroxisomes by bioinformatics analyses. Plant Physiol 135:783–800
16. Brocard C, Hartig A (2006) Peroxisome targeting signal 1: is it really a simple tripeptide? Biochim Biophys Acta 1763:1565–1573
17. Lametschwandtner G, Brocard C, Fransen M et al (1998) The difference in recognition of terminal tripeptides as peroxisomal targeting signal I between yeast and human is due to different affinities of their receptor Pex5p to the cognate signal and to residues adjacent to it. J Biol Chem 273:33635–33643
18. Lingner T, Kataya AR, Antonicelli GE et al (2011) Identification of novel plant peroxisomal targeting signals by a combination of machine learning methods and in vivo subcellular targeting analyses. Plant Cell 23:1556–1572
19. Reumann S, Chowdhary G (2018) Prediction of peroxisomal matrix proteins in plants. Subcell Biochem 89:125–138
20. Deng Q, Li H, Feng Y et al (2022) Defining upstream enhancing and inhibiting sequence patterns for plant peroxisome targeting signal type 1 using large-scale in silico and in vivo analyses. Plant J 111:567–582
21. Bongcam V, Petétot JMDC, Mittendorf V et al (2000) Importance of sequences adjacent to the terminal tripeptide in the import of a peroxisomal Candida tropicalis protein in plant peroxisomes. Planta 211:150–157
22. Chowdhary G, Kataya ARA, Lingner T et al (2012) Non-canonical peroxisome targeting signals: identification of novel PTS1 tripeptides and characterization of enhancer elements by computational permutation analysis. BMC Plant Biol 12:1–14
23. Kragler F, Lametschwandtner G, Christmann J et al (1998) Identification and analysis of the plant peroxisomal targeting signal 1 receptor NtPEX5. Proc Natl Acad Sci U S A 95: 13336–13341
24. Ma C, Reumann S (2008) Improved prediction of peroxisomal PTS1 proteins from genome sequences based on experimental subcellular targeting analyses as exemplified for protein kinases from Arabidopsis. J Exp Bot 59:3767–3779
25. Neuberger G, Maurer-Stroh S, Eisenhaber B et al (2003) Motif refinement of the peroxisomal targeting signal 1 and evaluation of taxon-specific differences. J Mol Biol 328: 567–579
26. Goodstein DM, Shu S, Howson R et al (2012) Phytozome: a comparative platform for green plant genomics. Nucleic Acids Res 40:1178–1186
27. Bolser D, Staines DM, Pritchard E et al (2016) Ensembl plants: integrating tools for visualizing, mining, and analyzing plant genomics data. Methods Mol Biol 1533:1–31
28. Weigel D, Glazebrook J (2006) Transformation of agrobacterium using the freeze-thaw method. CSH Protoc 2006:pdb.prot4666
29. Buchfink B, Reuter K, Drost HG (2021) Sensitive protein alignments at tree-of-life scale using DIAMOND. Nat Methods 18:366–368
30. Kullback S, Leibler RA (1951) On Information and Sufficiency. Ann Math Stat 22:79–86
31. Wagih O (2017) Ggseqlogo: a versatile R package for drawing sequence logos. Bioinformatics 33:3645–3647

Index

Berkley J. Walker (ed.), *Photorespiration: Methods and Protocols*, Methods in Molecular Biology, vol. 2792, https://doi.org/10.1007/978-1-0716-3802-6,

MIX
Papier aus verantwortungsvollen Quellen
Paper from responsible sources
FSC® C105338

If you have any concerns about our products,
you can contact us on
ProductSafety@springernature.com

In case Publisher is established outside the EU,
the EU authorized representative is:
Springer Nature Customer Service Center GmbH
Europaplatz 3, 69115 Heidelberg, Germany

Printed by Libri Plureos GmbH
in Hamburg, Germany